Cultivating Commerce

Sarah Easterby-Smith rewrites the histories of botany and horticulture from the perspectives of plant merchants who sold botanical specimens in the decades around 1800. These merchants were not professional botanists, nor were they the social equals of refined amateurs of botany. Nevertheless, they participated in Enlightenment scholarly networks, acting as intermediaries who communicated information and specimens. Thanks to their practical expertise, they also became sources of new knowledge in their own right. *Cultivating Commerce* argues that these merchants made essential contributions to botanical history, although their relatively humble status means that their contributions have received little sustained attention to date. Exploring how the expert nurseryman emerged as a new social figure in Britain and France, and examining what happened to the elitist, masculine culture of amateur botany when confronted by expanding public participation, Easterby-Smith sheds fresh light on the evolution of transnational Enlightenment networks during the Age of Revolutions.

SARAH EASTERBY-SMITH is Lecturer in Modern History and Director of the Centre for French History and Culture at the University of St Andrews. She has a PhD in History from the University of Warwick and has held postdoctoral fellowships at the Institute of Advanced Study, University of Warwick, the European University Institute and the Henry E. Huntington Library, California. She has served on the Executive Committee of the Social History Society (UK) and is a member of the British Society for Eighteenth-Century Studies, the British Society for the History of Science and the Society for the Study of French History.

SCIENCE IN HISTORY

Series Editors
Simon J. Schaffer, University of Cambridge
James A. Secord, University of Cambridge

Science in History is a major series of ambitious books on the history of the sciences from the mid-eighteenth century through the mid-twentieth century, highlighting work that interprets the sciences from perspectives drawn from across the discipline of history. The focus on the major epoch of global economic, industrial, and social transformations is intended to encourage the use of sophisticated historical models to make sense of the ways in which the sciences have developed and changed. The series encourages the exploration of a wide range of scientific traditions and the interrelations between them. It particularly welcomes work that takes seriously the material practices of the sciences and is broad in geographical scope.

Cultivating Commerce

Cultures of Botany in Britain and France, 1760–1815

Sarah Easterby-Smith

University of St Andrews, Scotland

CAMBRIDGE
UNIVERSITY PRESS

CAMBRIDGE
UNIVERSITY PRESS

University Printing House, Cambridge CB2 8BS, United Kingdom

One Liberty Plaza, 20th Floor, New York, NY 10006, USA

477 Williamstown Road, Port Melbourne, VIC 3207, Australia

314-321, 3rd Floor, Plot 3, Splendor Forum, Jasola District Centre, New Delhi - 110025, India

79 Anson Road, #06-04/06, Singapore 079906

Cambridge University Press is part of the University of Cambridge.

It furthers the University's mission by disseminating knowledge in the pursuit of
education, learning and research at the highest international levels of excellence.

www.cambridge.org
Information on this title: www.cambridge.org/9781107565685
DOI: 10.1017/9781316411339

First published 2018
First paperback edition 2019

A catalogue record for this publication is available from the British Library

ISBN 978-1-107-12684-8 Hardback
ISBN 978-1-107-56568-5 Paperback

To the memory of
Aline Torday, Jacqueline Fulton
and Marjorie Easterby-Smith.
Scholars, gardeners and inspirations.

Contents

Figures

Maps

Acknowledgements

This book has taken me much longer to complete than I would have liked, and I have accrued a great many debts along the way. The book started life as a doctoral thesis at the University of Warwick, and my thanks go firstly to my PhD supervisors, Maxine Berg and Colin Jones, whose teaching, friendly encouragement and rigorous reading of my work significantly shaped my approach to researching and writing history. I am indebted to many other people from my Warwick days, but in particular to Margot Finn and Stéphane Van Damme, and four of my fellow doctoral students: Katherine Foxhall, Philippa Hubbard, Michelle DiMeo and Kate Smith. Our wide-ranging discussions were immensely formative, and I still draw fruits from them now.

Emma Spary has been unstintingly generous in offering guidance, especially during my PhD, and the following have been equally liberal with advice or assistance at critical moments in the project's long evolution: Fredrik Albritton Jonsson, Bruno Belhoste, Marie-Noëlle Bourguet, Huw Bowen, Clare Brant, Margaret Carlyle, Neil Chambers, Pierre-Yves Lacour, Elise Lipkowitz, Daniel Kevles, Jim Livesey, Kapil Raj, Lissa Roberts, George Rousseau and Anne Secord. Conversations with Anne Goldgar, plus collaborative work with Emily Senior and others, including Alix Cooper, Helen Cowie, Florence Grant, Kathryn Gray, Judith Hawley, Gregory Lynall, Alexander Wragge-Morley and Anya Zilberstein, greatly influenced the final form taken by this book. At the European University Institute and then the Huntington Library, I found myself surrounded by two stellar groups of scholars, including Jorge Flores and Antonella Romano at the EUI and Daniela Bleichmar, Alex Marr and Neil Safier in Los Angeles. Also in LA, Ted Bosley, Zirwat Chowdhury, Elizabeth Eger (and her wonderful family), Steve Hindle, Lynn Hunt, Margaret Jacob, Heather James, Peter Mancall, Lindsay O'Neill, Roy Ritchie and Robert Westman all showed tremendous kindness and hospitality, and great generosity in answering my many questions. Thank you to *all* my fellow fellows at each of those institutions, who really epitomised the meaning of scholarly commerce.

I have been equally fortunate in joining an outstanding group of colleagues at St Andrews, many of whom, including David Allan, Riccardo Bavaj, John Clark, Aileen Fyfe, Colin Kidd, Konrad Lawson, Heidi Mehrkens (now at the University of Aberdeen), Guy Rowlands, Bernhard Struck and Richard Whatmore, have offered helpful insights on aspects of my research. Many thanks also to my past and present students, especially Dawn Hollis, Ramsay Mackenzie-Dodds, Paul Moorhouse, Elena Romero-Passerin and Matthew Ylitalo. Members of the Institute of Transnational and Spatial History read and discussed parts of the final draft of the manuscript, as did Kelsey Jackson Williams (now at the University of Stirling). Working with you all has been – and is – a delight, and I thank you for your patience with my questions, and for your critical engagement with my work.

The Economic and Social Research Council and the *Entente Cordiale* scholarship scheme generously funded my doctoral research. Subsequently, fellowships at the EUI, the Lewis Walpole Library and the Huntington Library afforded me the time and resources to work on this and other postdoctoral projects. The University of St Andrews has supported research trips to France, granted me a sabbatical to complete the final revisions and contributed towards the cost of image permissions and maps. I am also deeply grateful to the staff of the numerous libraries and archives who helped me plumb the depths of their holdings in search of obscure individuals. Michael Athanson of Orthodrome Geospatial made the maps for me, patiently working with the homemade sketches I sent him.

One of my greatest debts is to Cambridge University Press. Two anonymous readers carefully and rigorously considered my manuscript, as did Series Editors Simon Schaffer and Jim Secord, and I am grateful to all for their thoughtful suggestions. Lucy Rhymer's guidance and enthusiasm has been absolutely invaluable in enabling the manuscript to make its definitive move into print.

Special thanks go to Naomi Appleton, Catherine Clark, Brian Jacobson and Elsje van Kessel, who have offered friendships of the deepest and most meaningful kind – Catherine's comments on final drafts have also been especially helpful. Matthew Jarron held my hand through the book's final overhaul, and conjured up a delightful range of diversions for our time off together. The last vote of thanks, however, goes to my family, without whose love and support this book would not exist. My parents, Penny Summerfield and Mark Easterby-Smith, and my step-parents, Anna Lorbiecki and Oliver Fulton, inspired me to forge my own path as an academic, and have cheered the book on at every stage. Oliver and Penny also hold the dubious honour of having read and discussed more draft versions than anyone else – although any remaining errors are my own. I am deeply indebted to you all.

Note on the Text

Unless otherwise stated, all translations from French are my own.

Currency

25 French *livres* = approx. £1 sterling (until 1789).

Measures of Distance

1 French *league* = 2400 *toises*   Approx. 3 miles.
25 *leagues* = 1 *degré*   Approx. 60 miles.

Abbreviations

AD	Archives départementales, Loire-Atlantique
AN	Archives Nationales de France
AP	Archives de Paris
BL	British Library
BM	British Museum
BNF	Bibliothèque Nationale de France
Bodleian	Bodleian Library, Oxford
CRO	Cumbria Record Office, Carlisle
DTC	Dawson Turner Copies, Natural History Museum, London
JJ	John Johnson Catalogue of Printed Ephemera, Bodelian Library, Oxford
LC	Library of Congress
LRO	Lancashire Record Office
LS	Linnaean Society of London, Linnaean Correspondence Collection
MNHN	Muséum National d'Histoire Naturelle, Paris
NHM	Natural History Museum, London
OBP	Old Bailey Proceedings Online
PSJB	The Papers of Sir Joseph Banks Online, State Library of New South Wales
RBG	Royal Botanic Gardens, Kew
SV	Société Vilmorin, La Menitré, France
SVEC	Studies on Voltaire and the Eighteenth Century
TNA	The National Archives, London
WCRO	Warwickshire County Record Office
WLA	Warrington Library Archives

Introduction
Cultivating Commerce

In October 1776, the London nurseryman James Lee (1715–1795) (see Figure 1.1) wrote with excitement to the great Swedish botanist Carl Linnaeus (1707–1778):

I have this summer raised many new species of Mesembryanthemum, those that have flowered my daughter has delineated, if it would give you any pleasure I would willingly lend you the Drawings if you will describe them & send them back to me.[1]

Lee's enthusiasm about the *Mesembryanthemum* was almost tangible: these South African succulents were a relatively new discovery, and he was one of the first plant collectors in Britain to receive the specimens from the Cape of Good Hope. Lee was a nurseryman who gained his income by trading rare and unusual plants, and he undoubtedly hoped to make a profit from the sale of these small tender flowers.

James Lee did not, however, mention his pecuniary interests to Linnaeus. Instead, his letter mostly contained news about common acquaintances. 'I am charged by my Friend Francis Masson,' he explained, 'to send you the inclosed [sic] specimen & description of a plant that he has found in the Island of Madeira. He is desirous you would give it the name of Aitonia in honour to his Friend & patron Mr Aiton[,] Botanick Gardiner at Kew.' Lee then described the work of the British naturalist and patron of science Joseph Banks (1743–1820) and his assistant, the Swedish Linnaean Daniel Solander (1733–1782). The pair were busy describing and classifying specimens that they had received via a global network of collectors: 'Mr Banks' Herbarium is certainly the greatest & I believe the best that ever was collected. it [sic] is the daily labour of many servants to paste them [dried plants] on paper, And Banks & Solander spend 4 or 5 hours every day in describing and arranging them.'[2] In

---

[1] LS, L5238, James Lee (Hammersmith) to Carl Linnaeus (Uppsala), 4 October 1776.

[2] LS, L5238, James Lee (Hammersmith) to Carl Linnaeus (Uppsala), 4 October 1776. A herbarium is a collection of plants that have been pressed, dried and arranged according to a botanical system.

conveying news about other scholars of botany to Linnaeus, Lee sought to show that he was thoroughly integrated within a community devoted to the collection and study of the natural world. His associates ranged in status from gardeners to baronets, and they lived all over the world.

By the late eighteenth century, a number of commercial nursery gardeners like Lee had become differentiated from other plant traders, seed-sellers and florists. These 'elite' merchants focused on the collection and cultivation of outlandish ornamental plants newly imported to Europe. James Lee and his business partner Lewis Kennedy were reputedly leading men in London, and Lee especially acted as a liaison for traders, botanists and gardeners who wished to communicate with high-ranking men of science such as Linnaeus.[3] A small number of other plant merchants also rose to commercial and intellectual prominence in Britain and elsewhere in Europe. In Paris, for example, a nursery known by the names of its proprietors, Adélaïde d'Andrieux and Philippe-Victoire Lévêque de Vilmorin, had come to occupy an equally elevated position.

What distinguished these companies from other plant traders and gardeners was that they developed a sophisticated intellectual understanding of the science of plants and combined this with a green-fingered practical knowledge about their cultivation. This combination enhanced their ability to cultivate new plants and, by extension, offered them entry into scholarly networks like the one just sketched out. Linnaeus, Kennedy and Lee, and Andrieux and Vilmorin, differed in nationality, social status and occupation. Linnaeus had a university education and had been ennobled in 1762; the others were all gardeners who ran commercial plant nurseries. They spoke different mother tongues – Swedish, English, French – and adhered to different Christian denominations. Nevertheless, they exchanged letters, plant specimens, botanical paintings and books for over a decade. The plant traders specialised not only in raising rare plants but also in cultivating connections within the highly refined Enlightenment intellectual community.

Cultivating Commerce exposes and explores the roles that upper-end plant traders and gardeners played within transnational Enlightenment networks. Focusing on Britain and France, it reveals the wide range of connections that they forged and maintained. By linking scholars to a wider public of amateur botanists, certain traders and gardeners acted as conduits for knowledge that was variously practical, scientific or social.

---

[3] On Lee as a 'leading man', see: NHM, Department of Botany, Banks Correspondence, ff. 144–145, Joseph Banks (London) to Marmaduke Tunstall (Wycliffe, Yorkshire), 19 February 1786.

Equally, by promoting the scientific study of ornamental plants to a public apparently eager to purchase (and possibly to learn) botany, they contributed to framing the science as a cultural pursuit. The relatively humble status of most of these individuals, however, means that their contributions to science and to British and French culture have received little sustained attention to date.[4]

James Lee, Lewis Kennedy, Adélaïde d'Andrieux and Philippe-Victoire Lévêque de Vilmorin lived and worked between 1760 and 1815, a period characterised by immensely exciting developments in the history of botany, but also by extreme political turbulence. Prior to 1760, the study of botany in Britain had been in a state that might at best be described as moribund.[5] The decades that followed saw a seismic shift in the value placed upon botany, both by scholars and by the wider public. The multiple reasons for this rise in public estimation are explored in this book, but one of the most remarkable is the impact that voyages of exploration had on public interest in science. Between 1768 and 1771, for example, Captain James Cook's *Endeavour* expedition circumnavigated the world, mapping (and laying claim to) numerous new territories, including what became known as Australia. Over a thousand new species of plant arrived in Britain from the *Endeavour* voyage alone, and the presence of a huge diversity of exotic plants significantly rejuvenated public interest in botany.[6]

The end of the Seven Years' War in 1763 had decisively realigned the balance of French and British colonial power, redefining the areas from which plant hunters could obtain specimens and the routes that they could use to transport their precious charges. Large numbers of exotic, ornamental plants continued to arrive in both Britain and France. The tender specimens required careful cultivation by expert gardeners, and in France in 1764 André Thouin (1747–1824) was appointed to the post of

---

[4] The best existing work that situates British eighteenth-century plant nurseries within their cultural and scientific contexts is Coulton, 'Curiosity, commerce, and conversation'. Garden historian John H. Harvey undertook comprehensive research into the history of British (mostly English) plant nurseries in the 1970s and 1980s. Harvey's work offered an essential starting point for my own research, although he did not explore the plant trade from a cultural and scientific perspective. See especially: Harvey, *Early Gardening Catalogues*, *Early Horticultural Catalogues* and *Early Nurserymen* and Galpine, *The Georgian Garden*. Research into the French eighteenth-century plant trade is less well developed. For Paris, start with: Traversat, 'Les pépinières'.

[5] Thomas Martyn to James E. Smith, 16 November 1821, quoted in Shteir, *Cultivating Women*, p. 18. See also: Smith, 'A review of the modern state of botany' (1824), p. 386.

[6] In addition to the plants, naturalists on the *Endeavour* brought back more than 500 fishes, 500 bird skins, 1300 drawings and paintings, thousands of insects and several hundred ethnographic objects. Drayton, *Nature's Government*, p. 67. On the *Endeavour* voyage and its natural history dividends, see also: Banks *et al.*, *Sir Joseph Banks*; Beaglehole, *Life of Captain James Cook*; Carter, *Sir Joseph Banks*; Lincoln, *Science and Exploration*.

head gardener at the Jardin du Roi (the royal botanical garden) in Paris. Working with the Jardin's intendant, Georges-Louis Leclerc, comte de Buffon (1707–1788), and with the professors affiliated with the Jardin, Thouin helped to expand and transform the existing garden, bolstering its position as a primary destination for newly discovered plants.[7] As E. C. Spary has underlined, Thouin's personal participation in the international botanical network was essential to the garden's transformation.[8] *Cultivating Commerce* pieces together the histories of some of the other people who, although not employed at such prestigious institutions, were important associates of individuals such as Thouin.

The year 1789 is often used as a cut-off point in French history, but this would be an artificial end for the story told here. Within the botanical and horticultural world, the early years of the French Revolution were actually marked more by continuities than by decisive rupture. Many amateurs of botany continued to care for their gardens and specimen collections as best they could. 'My cultivations have not been interrupted by the revolution thanks to the good way of thinking of the majority of the habitants of the Boulonnais . . . I have nothing to complain about the new order of things', wrote the baron Georges-Louis-Marie Dumont de Courset from Normandy to Joseph Banks in December 1790.[9] British plant collectors likewise reported little change during the early 1790s, and continued to correspond with, and visit, their French counterparts, upholding a commitment to share precious scientific information regardless of the political context.[10]

Further revolutionary turmoil in 1793–94, however, swamped all but the most resilient horticulturalists.[11] Government by 'Terror' in France was accompanied by the outbreak of war, a development that ultimately obstructed botanical collecting networks and constricted the circulation of knowledge. Several notable savants were executed, including Chrétien-Guillaume de Lamoignon de Malesherbes (1721–1794), Louis XVI's loyal minister and a longstanding patron of French botany and gardening. More positively, the revolutionary government also refashioned the Jardin du Roi into the first French national museum, the Muséum

---

[7] Laissus, 'Le Jardin du Roi'; Letouzey, *Le Jardin des Plantes*; Spary, *Utopia's Garden*.

[8] Spary, *Utopia's Garden*, ch. 2.

[9] BL, Add. Ms. 8097, f. 400v, Dumont de Courset (Château de Courset, par Boulogne) to Joseph Banks (London), 20 December 1790.

[10] The amateur of botany Richard Twiss, for example, travelled to Paris in 1792 to see its gardens, and subsequently published an account of his travels. Twiss, *Trip to Paris*, pp. 1–2.

[11] Gardener Thomas Blaikie described the impact of the Revolution upon Parisian gardens (and gardeners) between 1789 and 1792 in his journal, later published in: Blaikie, *Diary*, pp. 221–239.

National d'Histoire Naturelle, a move that confirmed the prominent status of natural history within French revolutionary culture and that preserved the lives and livelihoods of several key botanists and plant traders.[12]

The transnational relationships forged by the surviving botanists and traders continued to evolve in significant ways in the early years of the nineteenth century. The Napoleonic Wars were characterised by strengthening nationalism on both sides of the Channel, and this further jeopardised the cosmopolitan ethos that had structured Enlightenment botany. 1815 marks a 'soft' ending for the book, signifying the removal of a political regime that had profoundly altered both the circumstances through which plants could be transferred and how collectors might obtain them.

This study of the plant trade and botany is inspired by a growing body of research that seeks to situate the history of science within its wider cultural and social context.[13] While 'science' may refer generically to a systematic, investigative approach to understanding the world, the purpose and content of such investigations have changed significantly over time. Early modern science differs strikingly from its present-day successor in terms of practice, participation and content. The boundaries between the strands of enquiry that we now think of as discrete disciplines were established gradually between the seventeenth and twentieth centuries. Eighteenth- and early nineteenth-century natural history is best conceived as a collection of practices and understandings that varied between – and were thus profoundly influenced by – the social and cultural contexts in which they emerged.[14]

Studying the activities and relationships formed by plant traders and gardeners offers a fruitful entry into understanding how botany was embedded within eighteenth- and early nineteenth-century society and culture. Both Britain and France saw a significant upturn in consumption in the eighteenth century: the middling ranks expanded, and more

---

[12] Spary, *Utopia's Garden*, ch. 5.
[13] For examples within this expanding field, start with: Bleichmar, *Visible Empire*; Clark et al., *The Sciences in Enlightened Europe*; Daston and Lunbeck, *Histories of Scientific Observation*; Daston and Pomata, *Faces of Nature*; George, *Botany*; Goldgar, *Tulipmania*; Jardine et al., *Cultures of Natural History*; Lynn, *Popular Science*; Miller and Reill, *Visions of Empire*; Parrish, *American Curiosity*; Safier, *Measuring the New World*; Shapin, *Social History of Truth*; Smith and Schmidt, *Making Knowledge*; Spary, *Utopia's Garden*; Terrall, *Catching Nature*.
[14] Easterby-Smith and Senior, 'Cultural production of natural knowledge'; Terrall, *Catching Nature*, p. 2.

people enjoyed a greater disposable income.[15] Certain individuals chose to invest their money and leisure time in collecting and arranging natural specimens. By the late eighteenth century, the study of natural history had become a fashionable pursuit for men and women in the middling and upper ranks, and it gained in social value.[16] Those amateur collectors who then communicated what they knew to others contributed to developing natural knowledge more generally.

Cultivating Commerce argues that public interest in the scientific study of plants stimulated new forms of social and economic commerce. It investigates how those commercial frameworks equally contributed to the evolution of scientific culture in the decades around 1800. The emerging trade in plants and other natural history specimens was closely linked to expanding public participation in botany, and botany, in turn, became ever more integrated within cultural and social life in France and Britain. The reciprocal relationships formed between traders, their clientele and their patrons impinged upon public participation in botanical science, influencing in particular ideas about the social status and gender of the botanical scholar. Those relationships also affected the selection of new exotic ornamental plants and the global networks through which those specimens travelled. Further, they substantially influenced the development of a new, associated science: horticulture.

Cultures of Botany

What exactly *was* botany in the later eighteenth and early nineteenth centuries? The scientific investigation of plants had been for centuries considered a branch of medicine, facilitating the study of *materia medica*, or drug specimens. Botany gained status as an independent area of enquiry in the seventeenth and eighteenth centuries: the words 'botanique' and 'botany' emerged in around 1611 in French and 1696 in English.[17] The science's move towards autonomy was connected more broadly to the

---

[15] On the shifting cultures of consumption between the seventeenth and nineteenth centuries, start with: Berg, *Luxury and Pleasure*; Berg and Clifford, *Consumers and Luxury*; Brewer and Porter, *Consumption*; Finn, *Character of Credit*; Glaisyer, *Culture of Commerce*; Roche, *History of Everyday Things* and *France in the Enlightenment*, ch. 5.

[16] The 'middling ranks' is a notoriously vague category that, in eighteenth-century Britain and France, could comprise anyone from an artisan to a gentleman. On the middling ranks, start with: French, 'Search for the "middle sort"'; Jones, 'Great chain of buying'; Maza, *Myth of the French Bourgeoisie*. On the eighteenth-century vogue for natural history, start with: Daston and Park, *Wonders*; Dietz and Nutz, 'Collections curieuses'; Drouin, 'L'histoire naturelle'; Jardine *et al.*, *Cultures of Natural History*; Laissus, 'Les cabinets d'histoire naturelle'; Pomian, *Collectionneurs*; Schnapper, *Le Géant*.

[17] 'Botanique' in *Le Trésor de la langue française informatisé*, http://atilf.atilf.fr/; 'Botany' in *OED Online*, http://dictionary.oed.com [accessed 3 May 2016].

development of an interest in taxonomy, or classification. All sorts of cultural and natural phenomena (ranging from the study of languages through to collections of books and naturalia) were labelled and then marshalled into hierarchies of increasing sophistication.[18]

One of the best-known Enlightenment taxonomic systems was that devised by Carl Linnaeus – described by James Lee in 1772 as the 'Father of Natural History'.[19] This regard was apposite: Linnaeus was widely acknowledged to be one of the eighteenth century's greatest authorities on botany. His method for classifying plants was taken up by scholars across Europe (and most especially in Britain) from mid-century onwards.[20] James Lee, in fact, was one of the first to publish an English explanation of Linnaeus' system, in his *Introduction to Botany* (1760).[21] Linnaeus' celebrated classificatory system was significant for its ease of use, as plants were ranked simply according to the number of sexual organs within the flower.

Linnaeus' other major contribution was to devise a binomial, or two-word, nomenclatural system. The cumbersome complexity of existing botanical names had previously restricted botanical study to well-educated aficionados, but Linnaean binomials offered for the first time an efficient, shorthand method of precisely describing specimens. The simplified names considerably eased communication among botanists and greatly facilitated public participation in botany. 'The passion for plants in this Country', Lee told Linnaeus in 1776, 'encreases [sic] every Day & I have the pleasure to tell you that your <u>sexual system</u> is more & more admired, & by none more than your affectionate Friend . . . James Lee'.[22]

Linnaeus' innovations were not without their critics, however: the most frequent charge levelled at his system was that it was awkwardly artificial, shoehorning specimens into a taxonomic arrangement that bore little relation to their actual state of being. The majority of Parisian botanists upheld the work of other systematisers, particularly those developing 'natural' methods of classification. The latter aimed to include all of a plant's characteristics within the classificatory schema,

[18] Cook, *Matters of Exchange*, pp. 25–28; Darnton, *Great Cat Massacre*, pp. 191–214; Gladstone, 'New world of English words', pp. 115–153.
[19] LS, L4741, James Lee (Hammersmith) to Carl Linnaeus (Uppsala), 23 October 1772.
[20] On Linnaeus, start with: Blunt, *Compleat Naturalist*; Delaporte, *Nature's Second Kingdom*; Duris, *Linné et la France*; Farber, *Finding Order in Nature*; Hoquet, *Fondaments de la Botanique*; Koerner, *Linnaeus*; Stafleu, *Linnaeus*.
[21] Lee's *Introduction to Botany* was a loose translation of Linnaeus' *Philosophia Botanica*. Henrey, *Botanical and Horticultural Literature*, vol. 2, p. 653; Shteir, *Cultivating Women*, p. 18.
[22] LS, L5238, James Lee (Hammersmith) to Carl Linnaeus (Uppsala), 4 October 1776.

seeking to arrange vegetables in ways more sympathetic to the complexity of the natural world. The French botanist Bernard de Jussieu developed what is now recognised as the first 'complete' natural system, laying out the Trianon garden at Versailles according to this new system in 1759 (details of which were later published by his nephew Antoine-Laurent in 1789).[23]

Rivalries certainly existed between the adherents of different systems, but the competing classifications did not prevent communication and exchange between botanists and plant traders. Carl Linnaeus and Bernard de Jussieu corresponded regularly while working on their contrasting systems, for example. Indeed, most botanists were adept at 'translating' taxonomic arrangements, and most, furthermore, were committed to ensuring that as few restrictions were placed on communication as possible. Enlightenment scholars ostensibly adhered to a 'cosmopolitan' philosophy that demanded cooperation in the interests of enhancing knowledge about nature.[24] James Lee's letters to Linnaeus were typical of most botanical correspondence, in that Lee relayed pieces of news that had originated from numerous continents. Defined by the worldwide movement of plants, people and new knowledge, eighteenth-century botany was a science that was global in content and transnational in structure.

Most historians have related botanists' global purviews to the specific national contexts from which they operated, and have emphasised the importance not only of taxonomy but also of agricultural improvement to eighteenth-century botanists. Joseph Banks' patronage of natural history has been framed within the 'English Enlightenment', and plant-collecting expeditions have been allied with the imperial aspirations of the British State.[25] Others have shown that the same was equally true for French botany, before and during the Revolution.[26] This book, which reconsiders the history of botany from the perspectives of plant traders, shifts the existing emphasis in two respects. Firstly,

---

[23] Delaporte, *Nature's Second Kingdom*; Duris, *Linné et la France*; Lawrence, *Adanson*; Stafleu, *Linnaeus*, ch. 9.

[24] This cosmopolitan commitment to free communication will be examined in Chapters 5 and 6. For further reading, start with: Crosland, *Scientific Institutions* and 'Anglo-Continental scientific relations'; Daston, 'Nationalism'.

[25] Gascoigne, *Joseph Banks* and *Science in the Service of Empire*. See also: Brockway, *Science and Colonial Expansion*; Damodaran *et al.*, *East India Company*; Drayton, *Nature's Government*; Grove, *Green Imperialism*; Harrison, 'Science and the British Empire'; Mackay, *In the Wake of Cook*; Schiebinger, *Plants and Empire*; Schiebinger and Swan, *Colonial Botany*.

[26] Bourguet and Bonneuil, *De l'inventaire du monde*; Bourguet *et al.*, *L'invention scientifique de la Méditerranée*; Lacour, *La République Naturaliste*; McClellan and Regourd, *Colonial Machine*; Spary, 'Peaches which the patriarchs lacked' and *Utopia's Garden*.

it shows that the cultivation of ornamental plants held a significant place within Enlightenment cultures of botany, as a means of drawing in public interest, and because gardens were ideal locations for experimenting on plants. Eighteenth-century botany was about much more than taxonomy and agricultural improvement. Secondly, it expands on the national perspectives offered by existing historical accounts, exploring the extent and nature of the connections forged between individuals in Britain and France. It places their experiences in the broader settings of transnational social relationships, comparative history and wider political developments.[27]

France and Britain share a long and somewhat uncomfortable history of mutual rivalry and admiration, which is as evident in commerce and science as it is in other areas of interaction.[28] Each country was a major player in the European Enlightenment in general, and in European botany more specifically. By the later eighteenth century, British scholars, especially those in London, had gained a distinguished place alongside counterparts in countries such as Sweden and the United Provinces (The Netherlands) as European leaders in both botany and horticulture.[29] On the other hand, French scholars, particularly those in Paris, were celebrated for their position at the 'centre' of the European Enlightenment, and likewise made significant contributions to botany. Institutions such as the Jardin du Roi and the Académie Royale des Sciences, as well as a host of private individuals (who people the subsequent chapters), conducted and shared extensive research into the workings of the natural world.[30]

Botanists and plant traders profited tremendously from cross-national exchanges during the eighteenth and early nineteenth centuries. Botanical study in Britain and France largely took place within a cultural context that was inherently cosmopolitan and characterised by close connections between commerce and science. While this book explores the characteristics of this cosmopolitan commercial context, it also

[27] I use the term 'transnational' to refer to the circulation of people, specimens and information across Europe, especially at a sub-national or non-state level. On approaches to transnational history and related methodologies such as *histoire croisée* and comparative history, start with: Cohen and O'Connor, *Comparison and History*; Saunier, *Transnational History*; Werner and Zimmermann, *De la Comparaison*.

[28] On Anglo-French relations in the eighteenth century, start with: Black, *Natural and Necessary Enemies*; Dziembowski, *Un Nouveau Patriotisme*; Morieux, *Une Mer Pour Deux Royaumes*; Tombs and Tombs, *That Sweet Enemy*. On cultural transfers and the circulation of knowledge, start with: Hilaire-Pérez, *L'Invention Technique*; Ogée, *Better in France?*; Rabier, *Fields of Expertise*; Thomson *et al.*, *Cultural Transfers*.

[29] Drayton, *Nature's Government*; Gascoigne, *Science in the Service of Empire*.

[30] On the perceived centrality of Paris to the Enlightenment start with: Belhoste, *Paris Savant*; Romano and Van Damme, 'Sciences et villes-mondes'; Van Damme, *Paris*.

problematises it by considering the extent to which national and local cultures influenced the scientific study of plants. Differences emerged, for example, between the constitution and reach of domestic plant exchange networks, among attitudes towards women's participation in both commerce and amateur botany, and with regards to how individual actors negotiated competing political loyalties. This book explores the extent – and the limits – of the cosmopolitan scholarly commerce in plants.

Science and its Publics

European Enlightenment culture was characterised by a notable upturn in inquisitiveness into the world. By the latter half of the eighteenth century, curiosity, instruction and educated conversation carried substantive social credit.[31] In particular, collecting was considered central to Enlightenment scholarship and was intimately linked to the expanding consumer culture. The *Wunderkammern*, or curiosity cabinets, of the sixteenth and seventeenth centuries formed core resources for early modern scholars, as did the gardens that surrounded them. Forming a collection inscribed an individual within a range of social practices that could in turn result in access to more specimens and more information. Participation, however, was until the mid-eighteenth century largely restricted to the social elite.[32]

Like their predecessors in earlier centuries, eighteenth-century scholars were still expected to possess a collection; they visited and wrote about one another's cabinets and gardens, and judged their counterparts according to the quality and value of the specimens they had collected and the tastefulness of their display.[33] Increasing disposable income and leisure time among the middling and upper ranks meant that the number of private, domestic collections increased and that the 'public' interested in learning about science broadened out socially – including, for the first time, significant numbers of women.[34]

[31] On Enlightenment culture, see references elsewhere in this Introduction and also: Goodman, *Republic of Letters*; Knott and Taylor, *Women, Gender and Enlightenment*; Roche, *France in the Enlightenment.*

[32] Brockliss, *Calvet's Web*; Daston and Park, *Wonders*; Findlen, *Possessing Nature*; Goldgar, *Tulipmania*; Guichard, *Les amateurs*, esp. chs 3 and 6; MacGregor, *Curiosity and Enlightenment*; Miller, 'Joseph Banks'; Pomian, *Collectionneurs.*

[33] Bleichmar, 'Learning to look'; Guichard, *Les amateurs*, ch. 4; Spary, 'Scientific symmetries'; van de Roemer, 'Neat nature'.

[34] By 'public', I mean especially the members of the middling and upper ranks who engaged in some way with the new opportunities to learn about science. For more on the emergence of public science during the Enlightenment, start with: Broman, 'Habermasian public sphere'; Sutton, *Science for a Polite Society*. See also the references in the next footnote.

Science became a significant part of public culture during the eighteenth century, especially for men and women of the middling and upper ranks. The forms in which scientific knowledge was communicated, however, varied according to the gender and social status of the recipient.[35] By the latter half of the eighteenth century, enterprising lecturers advertised courses to be delivered in lecture halls, gardens, museums and public houses, at fairs or in their own homes. While they could always rely on regularly attracting enthusiastic audiences, the choice of location determined what kind of person might attend.[36] New kinds of museums also emerged, ranging from multiple, relatively small-scale provincial enterprises with a clear commercial purpose to the first major national museums.[37] The collections in both the British Museum, which opened its doors to the public in 1759, and France's Muséum National d'Histoire Naturelle, founded 1793, were explicitly held in trust for their respective nations.[38] Prospective visitors might gain admission by applying for a ticket. The taste for learning among the middling and upper ranks both nurtured and was fed by an emerging educational industry, which made knowledge interesting and accessible via a whole host of commodities, including books, instruments and even pre-prepared collections.[39] Merchants and educators, including the plant traders discussed here, were key to the dissemination of Enlightenment culture. The Enlightenment reciprocated by offering them unrivalled commercial and social opportunities.

To what extent did the different social groups who purchased plants or visited museums or botanical gardens actually interact with one another? And what did it mean for a middle- or upper-ranking woman to assert an interest in science? The upper-end nursery traders and gardeners discussed here offer a distinctive perspective on public participation in science. As go-betweens, they negotiated relationships with the very different groups interested in collecting and studying plants, groups that varied widely in terms of their backgrounds and levels of knowledge. The traders forged and sustained social and intellectual connections; as a result of their position at the hub of several different kinds of knowledge networks, they also became significant sources of new knowledge in their own right. The ways in which these socially diverse groups interacted, as

[35] Golinski, *Science as Public Culture*; Lynn, *Popular Science*; Stewart, *Rise of Public Science*.

[36] Public houses, for example, primarily attracted male audiences. On the theatrical aspects of French science demonstration lectures and developing notions about the French and British publics as spectators, see: Camp, *The First Frame*, ch. 3; Gillespie, 'Ballooning'.

[37] Brears, 'Commercial museums'; Haynes, 'A "natural exhibitioner"'; Torrens, 'Natural history'.

[38] Anderson *et al.*, *Enlightening the British*; Spary, 'Forging nature' and *Utopia's Garden*.

[39] Lynn, *Popular Science*; Stewart, *Rise of Public Science*; Sutton, *Science for a Polite Society*.

well as the cultural connotations of their participation, are elaborated throughout the course of this book.

The taxonomic advances outlined earlier represent huge milestones in the history of botany overall. We will see, however, that the 'botany' that attracted public attention in the late eighteenth and early nineteenth centuries was characterised by a wider range of goals and practices than taxonomy alone – despite the fact that the eighteenth century has been characterised as the 'age of classification'.[40] When viewed from the perspectives of the plant traders and the people who they served, the practical application of botanical knowledge in gardens was just as important as arranging plants systematically.

Horticulture occupied a central place, alongside medicine, agriculture and industry, as a utilitarian destination for botanical expertise. A glance through the articles on natural history in the great Enlightenment dictionary, Diderot and D'Alembert's *Encyclopédie*, confirms this less restrictive interpretation. In his discussion of the relations between the different forms of knowledge, for example, Denis Diderot emphasised that botany included the study of plants' 'economy, propagation, culture, vegetation, etc', and he classed both 'agriculture' and 'gardening'.as 'branches' of the science.[41]

Cultivating one's garden was not just an end-point for botanical learning. It was also an arena within which further knowledge might develop. Some of the earliest practical investigations into plant growth and physiology, for example, were undertaken by gardeners. We will see that plant traders like Kennedy and Lee, Andrieux and Vilmorin and their associates acted as sources of and conduits for such information, translating and transmitting their findings to their scholarly counterparts. This history of experimentation and practical learning, however, was often omitted from contemporary botanical records.

Enlightenment cultural, social and institutional structures meant that the world inhabited by the traders and the groups they served was quite different from that which we know today. The word 'scientist' did not yet exist, for example, and the clear differentiation between 'professional' and 'amateur' scholars emerged only in the late nineteenth century, a product of the institutionalisation of scientific research and the development of qualifications that defined expertise.[42]

Although botany was taught widely in Enlightenment Europe, it was not possible, for example, to obtain a university degree in the subject. As a consequence, the meanings of terms such as 'botanist', 'botanophile'

---

[40] Foucault, *Order of Things*, ch. 5. See also: Hooper-Greenhill, *Museums*.

[41] Diderot, 'Explication détaillée'. See also: Anon., 'Jardinage', p. 460. On the emergence of horticulture as a scientific discipline, see: Lustig, 'Creation and uses of horticulture'.

[42] Alberti, 'Amateurs and professionals'; Fyfe and Lightman, *Science in the Marketplace*.

and 'amateur' were marked by considerable fluidity. A very small number of individuals were actually employed as 'botanists' by institutions or private patrons, but people lacking any formal status whatsoever might also claim this identity. The terms 'botanophile' and 'amateur of botany' could pejoratively signify someone who understood little or nothing of the plant sciences, but could alternatively denote someone who garnered high respect for their scholarship. The Société des Botanophiles in the French town of Angers was an esteemed intellectual group, for example; in neighbouring Nantes, the author of a report lobbying for the creation of a botanical professorship in 1795 similarly used the term 'Botanophile' to assert his own knowledge and understanding.[43]

This vagueness of language might suggest that scientific participation was open to anyone. But each label in fact encoded a range of cultural expectations about how scholarship should be undertaken and, therefore, who might partake in it. In particular, the notion of being an 'amateur' of botany conveyed a host of unwritten assumptions about the aesthetic sensibilities and cultured refinement of the practitioner, assumptions that were also linked to ideas about gender and social status. The ways in which amateur scholars from a range of backgrounds engaged with the 'polite' and sociable commerce of natural history are explored later in the book. From the start, however, it is important to underline that the term 'amateur' did not – indeed, *could* not – simply signify the opposite of 'professional'.

Despite the ambiguity in the terminology, Enlightenment intellectual communities remained strongly hierarchical. Most scholars in the eighteenth century were members of what was known as the 'Republic of Letters', an international network of intellectual individuals (mostly, but not exclusively, men) who corresponded – often across huge distances – on a range of topics.[44] Their letters (which could extend to lengthy manuscript dissertations) were then circulated and discussed among local circles of scholars; the fruits of this correspondence and conversation came to comprise much of what we now think of as Enlightenment knowledge. The late eighteenth-century Republic of Letters was important to science in general and to botany in particular, for it provided the medium through which knowledge was constructed and exchanged.

---

[43] AD, L626, Observations relatives à l'établissement projeté d'un jardin de Botanique en la commune de Nantes, 15 Gérminal Year 3 [4 April 1795].

[44] Brockliss, *Calvet's Web*, pp. 13–14; Goodman, *Republic of Letters*; Hahn, *Anatomy*, ch. 2. Some historians have preferred to use 'Republic of Science', but this term overemphasises disciplinary specialism. Enlightenment scholarly culture was primarily characterised by scholars' multiple intellectual interests. For 'Republic of Science', start with: de Jouvenel, 'The Republic of Science'; Hagstrom, *Scientific Community*.

While letter writing was perhaps a defining feature of Enlightenment scholarship, travel was equally essential. The act of making new acquaintances allowed scholars to construct and maintain their reputations; a new connection could increase access to the manuscripts and objects from which new knowledge might be forged.[45] The British connoisseur William Constable (1712–1791), for example, took three Grand Tours through Continental Europe during his lifetime, latterly with his sister Winifred in 1769–71. The scholars they called on included prominent luminaries such as Jean-Jacques Rousseau, and Constable subsequently maintained regular correspondence with the people they met.[46] In return for the information and objects Constable sent his new contacts, they furnished him with materials that ultimately came to make up a majestic collection. Constable was an archetypical connoisseur, whose interests encompassed (among other subjects) experimental philosophy (especially electricity and chemistry), antiquities, zoology and botany.[47]

Formal recognition of intellectual merit could be granted through election to an academy or learned society. William Constable never specialised in a particular subject, but nevertheless was made a fellow of the Society of Antiquaries and the Royal Society.[48] Competition for election to top institutions such as the Royal Society and Académie des Sciences was intense, however. Successful election was not just a guarantee of intellectual proficiency: it also brought further social connections, scholarly support and, exceptionally for full members of the Académie des Sciences, a salary.[49]

Joseph Banks followed a similar model of connoisseurship to that of William Constable, taking his own 'grand tour' around the world and maintaining an interest in numerous branches of knowledge, albeit with a clear concentration in natural history.[50] Chrétien-Guillaume de Lamoignon de Malesherbes, mentioned earlier in the context of the French Revolution, likewise created an extensive collection at great personal expense and exchanged seeds, plants and practical information with other collectors and traders.[51] Both Malesherbes and Banks accrued

---

[45] Goldgar, *Impolite Learning*; Meredith, 'Friendship and knowledge', pp. 151–191.

[46] Connell, 'Grand Tour'; Credland, 'Introduction'; Hall, 'Cabinet of scientific instruments'.

[47] Hall and Hall, *Burton Constable Hall*, pp. 25–26, 33, 42–43, 48–50.

[48] Constable was proposed for election to the Royal Society in 1774 by Daniel Solander and Joseph Banks. Hall, 'Cabinet of scientific instruments', p. 31. Dual membership of both societies was very common; many gentlemen virtuosi consequently used the initials 'FRASS'. Gascoigne, *Joseph Banks*, p. 60.

[49] Hahn, *Anatomy*, esp. chs 2 and 4; Lyons, *Royal Society*, chs V and VI.

[50] Gascoigne, *Joseph Banks*, pp. 60–61.

[51] Malesherbes' botanical notes and correspondence are now held in AN, 399/AP/97 – 399/AP/101. For more on Malesherbes as a scholar and his connections with other

numerous academic fellowships, which for Banks culminated with election as President of the Royal Society of London in 1778.

Membership of the Republic of Letters expanded during the eighteenth century, in parallel with growing participation in the Enlightenment. Nevertheless, interactions between scholars continued to be structured as they had been in the seventeenth century.[52] Although it was theoretically possible to reach the top of the tree through intellectual contributions alone, social connections and financial resources counted hugely towards establishing a position within the scholarly hierarchy.

Cultivating Polite Science

It is perhaps not surprising, then, that very few plant traders and gardeners gained election to scientific societies.[53] For many, 'success' might be defined instead as obtaining regular access to rare, exotic plants, the subsequent sale of which could result in significant commercial dividends. Access to rare plants, however, depended on regular interaction with the members of that scholarly world. As we will see, men like James Lee and Philippe-Victoire Lévêque de Vilmorin became adept at conducting themselves in ways that allowed them to penetrate these scholarly circles. The performance of politeness was key to ensuring their successful integration.

'Politeness', a term often used in the context of British eighteenth-century culture, was understood to comprise sociability, refinement and gentility. Polite persons controlled their behaviour and emotional responses, showing their affinity with the highest social order. Politeness was contrasted with the four *faux pas* of vulgarity, rusticity, barbarity and utility; in France, the parallel cultural phenomenon was more often described as '*civilité*'. In both countries, it was the metropolitan social elite that sketched out the contours of polite or civil culture: although they formed a tiny minority, the upper ranks nevertheless wielded tremendous cultural influence.[54]

---

botanists, especially Buffon and Rousseau, start with: des Cars, *Malesherbes*; Gillispie, *Science and Polity in France at the End of the Old Regime*, pp. 9–10, 19–20; Roger, *Buffon*, pp. 194–195; Rousseau and Malesherbes, *Jean-Jacques Rousseau*; Williams, *Botanophilia*, pp. 51–54, 96–99.

[52] Brockliss, *Calvet's Web*, 'Introduction'; Goldgar, *Impolite Learning*.

[53] Notable exceptions include André Thouin, elected to the Académie des Sciences in 1786, and Richard Bradley (1688–1732), elected to the Royal Society in 1712.

[54] France, *Politeness*, esp. pp. 57–58; Greig, *Beau Monde*, p. 20; Klein, 'Politeness for plebes', p. 365 and 'Politeness and the interpretation'; Langford, *Polite and Commercial People*; Lilti, *World of Salons*; Tikanoja, *Transgressing Boundaries*.

Distinctions between and within the two national cultures meant that politeness and civility were defined according to slightly different standards. British commentators thus often characterised the French as foppish or effeminate; the French tended to portray polite Britons as stand-offish or graceless. There were also variations within each country among metropolitan and provincial understandings of good comportment. The gay sociability that characterised the burgeoning French public sphere after 1750, for example, was in fact completely the reverse of the culture of the royal court, where courtiers doggedly adhered to strictly hierarchical codes of conduct. In Paris, by contrast, behaviours such as 'kissing and handshaking' (which disgusted dour British observers) greatly facilitated interaction among different social groups.[55] Europe in the eighteenth century might certainly be characterised as 'polite', but beyond a shared understanding of basic rules of etiquette there were multiple ideas of what exactly politeness consisted of. Polite culture varied according to both social status and regional character.

The expansion of a consumer society in eighteenth-century Britain and France facilitated the promotion of polite culture and encouraged new forms of sociability.[56] Although politeness in Britain was equated with the culture of the upper social levels, modes of being polite were adopted, to a greater or lesser extent, by non-elite sections of society from the early eighteenth century onwards. 'Very useful manuals', encyclopaedic guides and even Joseph Addison's *Spectator* equipped readers with the necessary cultural literacy to adopt politeness as a behavioural norm.[57] Those whose occupations brought them into regular contact with members of socially superior ranks could use politeness or civility as a means of bridging the social gap.[58] Traders such as Lee and Vilmorin thus reflected a wider trend among the educated middling ranks in Britain and France.

Learning new forms of behaviour is not easy, however, and there was much variation in the extent to which the middling and lower ranks adopted forms of polite behaviour. Caricaturists and writers delighted

[55] Jones, *Smile Revolution*, pp. 91–92. See also: Reddy, *Navigation of Feeling*, pp. 122–130, 141–147.

[56] Berry, 'Polite consumption'; Coquery, *La boutique et la ville*; Klein, *Shaftesbury and the Culture of Politeness*, pp. 1–24; Stobart *et al.*, *Spaces of Consumption*.

[57] Klein, 'Politeness for plebes', p. 365 and 'An artisan in polite culture'.

[58] In using the word 'bridging', I do not mean to suggest that social differences were either lessened or encroached upon, however. Although some of the people discussed here eventually made substantive fortunes through trading plants, most remained securely – and deferentially – within their place. Most of the plant traders and gardeners discussed in this book sought to be seen as 'genteel', but not as gentlemen. On the construction of the 'genteel' or 'gentlemanlike' and its relationship to politeness, see: Klein, 'Politeness for plebes'; Vickery, *Gentleman's Daughter*, ch. 1.

in depicting the robust 'impolite' cultures that also persisted throughout the century.[59] This book explores the culture of politeness, and its limitations, from the perspectives of plant traders and gardeners. Most did not come from 'polite' backgrounds, but those who managed to assimilate and espouse the behavioural model developed among the higher social echelons generally enjoyed greater commercial and social success. Not all plant traders, gardeners and plant hunters were equally socially astute, however, as we will see.

Enlightenment science has also been described as 'polite'. Strange though it may now seem, it was conventional in an age before professional scientific qualifications to calibrate the credibility of an informant on the basis of their behaviour.[60] One needed to act 'respectably' (read: 'politely') in order to appear a reliable member of the scholarly community, and correct comportment was further connected to assumptions about one's moral state. The ability to communicate, either through oral conversation or in written form, was also central to such judgements. According to Bluestocking Hannah More, 'the noblest commerce of mankind' was 'conversation' because it would lead to the improvement of morals and civil virtue.[61] The incentives for plant traders and gardeners to assume forms of polite behaviour related to science as well as to commerce.

Knowing how to behave politely was thus important for plant traders who wished to make headway within scholarly networks. The conversation and commerce that they conducted in part defined the circulation of plants and the spread of knowledge. In 1726, the botanist and writer Richard Bradley encouraged fellow gardeners to engage in sociable commerce, for, he explained, 'Conversation promotes Experience, and Experience leads to Perfection.'[62] We will see how Bradley's invocations were interpreted over the century that followed. Sociable exchanges, which

[59] Carr, 'Polite and enlightened London?', p. 634; Coltman, *Classical Sculpture*; Jones et al., *Saint-Aubin*; Kelly, *Society of Dilettanti*.

[60] Shapin, *Social History of Truth* and 'A scholar and a gentleman'; Walters, 'Conversation pieces'.

[61] Hannah More, 'Bas Bleu, or Conversation. Addressed to Mrs Vesey' (1786), quoted in Eger, 'The noblest commerce', p. 289. The Bluestockings formed the most prominent circle of British scholarly women in the second half of the eighteenth century. The term 'Bluestocking' eventually came to denote any female writer or intellectual, but the label initially described a mixed group of men and women who attended the soirées of Elizabeth Montagu and Elizabeth Vesey in London and Bath from the 1750s onwards. Following the French salon model, the Bluestockings promoted a form of convivial intellectuality that was more widely accepted than that of the earlier, isolated, learned woman. On the Bluestockings, start with: Eger, *Bluestockings* and 'Luxury, industry and charity'; Myers, *The Bluestocking Circle*.

[62] Bradley, *General Treatise* (1726), p. 348. For more on Bradley and his world, see: Coulton, 'Curiosity, commerce, and conversation'.

were conducted either in person or through a 'commerce of letters', were fundamental both to the spread of botanical knowledge and to its cultural expression in the Age of Enlightenment.[63]

Piecing together the histories of nurseries such as those of Kennedy and Lee and of Andrieux and Vilmorin has been central to the task of situating botany within its broader social and cultural context. The history of the nursery trade, however, might perhaps be characterised more than anything by a paucity of sources. Only around twenty letters written to or from Lewis Kennedy and James Lee now exist, for example, in spite of the distinguished history of the company. Even less detailed information about Andrieux and Vilmorin's eighteenth-century activities has survived to the present day, because the Vilmorin family archive was destroyed in a fire in Paris in the mid-twentieth century.[64] This dearth posed a tantalising challenge initially, and this book brings together evidence now scattered across archives in Britain, France and North America.

The six chapters that follow examine the different groups that participated in plant collecting networks in Britain and France in the late eighteenth and early nineteenth centuries. Chapters 1 and 2 focus on the plant traders, recovering their histories from the available evidence and examining the importance of commercial networks to the circulation and study of plants. The nurseries at the top end of the market boosted their business prospects through the development of practical skills and by acquiring botanical knowledge. But success within the plant trade was not only a matter of growing plants: the upper-end nurseries also cultivated social expertise, forming strong relationships with their customers and correspondents. In turn, they came to occupy key positions as intermediaries within botanical and horticultural networks, and, as a result of their activities, eventually gained significant social and cultural distinction.

Who, however, were the nurseries' customers? And what did it mean, both culturally and socially, to study botany or collect plants? Chapter 3 argues that we need to consider the multiple ways in which people engaged with nature scientifically, and understand the consequences of that participation for ideas about masculinity and femininity. It examines how the study of botany became integrated within eighteenth-century culture, focusing on the figure of the scholarly amateur of botany. The

---

[63] Goldgar, *Impolite Learning*, p. 18.
[64] Sosthène de Vilmorin, personal communication 15 July 2008.

chapter – and the book as a whole – adopts an inclusive understanding of 'botany', aiming to capture the multiple ways in which people studied plants. Although the people who learnt how to use botanical systems might have described themselves as 'botanists', they did not necessarily apply their knowledge in ways that would now count as 'scientific' or 'botanical'.

The specimens that plant hunters sent back from their collecting missions not only led to the development of new botanical knowledge and further embellishment of gardens, but were also the bases on which reputations and livelihoods were constructed. Chapter 4 examines the extent to which collectors hunting for horticultural specimens participated in the Republic of Letters. It focuses on the example of a young Scottish plant hunter, Thomas Blaikie, who, although not considered a 'scholar', and in spite of his low social rank, became integrated within the Republic of Letters as a result of his travels. Blaikie's story helps to enhance our understanding of the social structures that supported the Republic of Letters in the later eighteenth century, showing them to be more diverse in form than previously thought.

Finally, Chapters 5 and 6 investigate how plant merchants and plant hunters balanced their occupational identities against their claims to be scholarly practitioners. Understanding the careful positioning undertaken by the merchants and their emissaries involves situating the plant trade within the context of the global circulation of knowledge, a context that was profoundly sensitive to broader political transitions. These chapters extend the book's scope geographically and chronologically, following the plant hunters across the seas to the colonies and trading posts they plundered, and tracking their activities into the early nineteenth century.

What methods did the traders and collectors use to negotiate the problems posed by rivalries and conflicts? Chapters 5 and 6 consider in particular how the participants in botanical networks (traders and scholars) deployed cosmopolitan philosophies to facilitate global exchange. Merchants, for example, had to square their private interests – their need to obtain and sell plants – with the broader, non-partisan ethos of Enlightenment science. The outbreak of revolution in France, and of war among European countries, threw such careful balancing into jeopardy. Plant collecting may have continued apace during the Age of Revolutions, but the tenets supporting cosmopolitan science were greatly destabilised by intensifying national conflicts.

The botanical networks discussed here were all in some way connected to Enlightenment institutions, but they were also much broader, spilling over into civil society and featuring participants who possessed a broad range of expertise and understanding. Together, their members crafted cultures of botanical learning that came to occupy a major place in British and French society by the turn of the nineteenth century.

1   Plant Traders and Expertise

The 1810 edition of James Lee's *Introduction to the Science of Botany* opened, unusually, with an engraving of its author (Figure 1.1).[1] The portrait, of which the colour original is now in the Hammersmith and Fulham Archives, depicts the nurseryman and gardener modestly bewigged and plainly dressed, closely examining a plant. Semi-profiled in front of a fiery sunset, we see Lee completely absorbed in studying the specimen, so dedicated to understanding its detail that he ignores the dramatic skies around him. The portrait blends together a scholarly, gentlemanly demeanour with Lee's more humble professional identity. His flushed, heavy-set features and his large, oversized hands intimate a world of work far removed from the everyday experiences of most British gentlemen. The shaft of light falls, however, on his forehead and leads down to his eyes and to a tiny magnifying glass: Lee, like a gentleman scholar, is engrossed in intelligently examining the specimen. No rough gardener after all, here we see a man of understanding whose whole character is subsumed in his commitment to botany.

Born to a lowly but 'respectable' family, James Lee (1715–1795) had walked from his native Selkirkshire to London in the early 1730s, a journey of approximately 350 miles, to learn the art of gardening.[2] Lee, like many of his associates, was not from a cultured background, and yet by the close of his life he was depicted as a quasi-gentleman. His reputation (as we saw in the Introduction) extended not just across Britain but around the world. His route to social elevation was not an obvious one, however. How did simple gardeners-turned-plant-traders manage to gain such merit?

---

[1] The engraving, by S. Freeman, was reproduced in the frontispiece to Lee, *Introduction to the Science of Botany* (1810).

[2] Very little is known about Lee's parents, although his biographer E. J. Willson speculates that they may have been linen weavers. Willson cites a 1795 edition of the *Gentleman's Magazine*, which describes them as 'respectable . . . but not in a station that allowed them to give him any farther education than is in the power of everyone to attain in that part of Britain [lowlands Scotland]'. Willson, *James Lee*, pp. 3, 1.

Figure 1.1 James Lee (1715–1795), from James Lee, *Introduction to the Science of Botany* (London, 1810), frontispiece. King's College London, Foyle Special Collections Library.

This book opened with a discussion of Lee's letter to Linnaeus of 1776, which introduced the overlapping worlds of Enlightenment enquiry, commerce and collecting. This chapter uncovers the social history of a small but significant group of plant traders and gardeners who, like Lee, worked within these worlds. It situates their activities within wider botanical and commercial contexts, seeking to understand how certain plant traders and gardeners achieved distinction. Lee, his French counterpart Vilmorin and their other associates presented something new to *both* botany and horticulture, something special, which attracted the attention of scholars as well as general consumers. Their history is one of practical skill and intelligence, and also of social astuteness and an ability to communicate. Most importantly, these traders developed a form of hybrid expertise – which related to dealing with other people as much as with plants.

Expertise and the Elite Nurseries

Who were the people who rose to the 'top' of the gardening profession? Several names occur with notable frequency in the correspondence exchanged between botanists. In London, these are individuals such as James Gordon, William Malcolm, William Curtis and partners James Lee and Lewis Kennedy. Parisian botanists, equally, conversed with a small number of collectors and traders: Germain Jouette, M. Mathieu, the apothecary Descemet and the plant traders Adélaïde d'Andrieux and Philippe-Victoire Lévêque de Vilmorin. These individuals all espoused the twin aims of increasing knowledge about botany and furthering their own prospects as collectors and traders of plants. They developed a specific expertise in the naturalisation and cultivation of unknown exotic plants – a particularly challenging task that required high levels of practical skill and an advanced understanding of plant growth and physiology. Although their names appear rarely in archival indexes, they recur time and again within the correspondence between important botanists whose letters, unlike most of the records kept by gardeners and traders, have been preserved and catalogued. Patrons of science and botanical authorities, including Joseph Banks, John Fothergill, Chrétien-Guillaume de Lamoignon de Malesherbes and André Thouin (on all of whom, more later) cited these same names repeatedly, mentioning them not only as suppliers of plants, but also as valuable sources of information and expertise.

By amplifying the visibility of their specialist expertise, these plant merchants were able to develop close connections with important

scholars. Yet where did they come from? The histories of our two principle companies, Kennedy and Lee in London and Andrieux and Vilmorin in Paris, reveal how these gardeners-turned-traders gained such
high acclaim. Their rise to prominence was exceptional compared to the
wider plant trade yet typical for the small number of merchants who
fused horticulture with an extensive botanical understanding. Almost all
the traders discussed here started life humbly as gardeners but ultimately
rose to participate in international Enlightenment networks.

Kennedy and Lee's Vineyard Nursery

The Vineyard Nursery enjoyed a particularly illustrious history. Founded
in 1745, it had become well known in British botanical and horticultural
circles by 1760.[3] Its proprietors, Lewis Kennedy (1721–1783) and James
Lee, forged and maintained strong connections with major botanists
across Europe, including Carl Linnaeus and Joseph Banks. As a consequence of these links, the Vineyard became one of the first to receive
exotic specimens brought back by plant hunters.[4] It remained a family
business until the 1890s: sons John Kennedy (1759–1842) and James
Lee Jr (1754–1824) inherited the nursery after their respective fathers'
deaths.[5] By the early nineteenth century, its reputation had extended
across Continental Europe, supplying customers of no lesser rank than
Empress Josephine of France (1763–1814)[6] and Maria Feodorovna,
Dowager Empress of Russia (1759–1828).[7]

In establishing the Vineyard Nursery, Kennedy and Lee followed a
precedent set by other head-gardeners-turned-nurserymen. Head gardener James Gordon (c. 1708–1780), for example, founded his famous
Mile End Nursery following the death of his employer Lord Petre in
1742.[8] Practical skills and botanical knowledge were central to the commercial success of such nurseries, and Lewis Kennedy and James Lee
both gained significant experience working in gardens in and around
London before founding their own establishment. Kennedy, like Lee,

---

[3] BL, Add. Ms. 35057, Rev. John Walker (Moffat) to J[oseph] Banks Esq. (London),
   28 March 1767; Faulkner, *History and Antiquities*, p. 43; Thornton, 'Sketch of the life of
   James Lee', pp. xiii–xix; Webber, *Early Horticulturalists*, p. 92.

[4] Andrews, *Botanists' Repository*, vol. 2, pl. 82 'Banksia Serrata'; Carter, *Sir Joseph Banks*,
   pp. 42, 160, 174, 282.

[5] Willson, *James Lee*, pp. 63–64.

[6] Bourret *et al.*, *Parcours*, pp. 192–196, 224–225; Hobhouse, *Plants in Garden History*,
   p. 220; Willson, 'Farming, market and nursery gardening', pp. 93–94 and *James Lee*,
   pp. 35, 54–55.

[7] BL, Add. Ms. 47237, Lieven Papers, ff. 15–16, 37–38, 81, 84; BL, Add. Ms. 47237,
   Lieven Papers, f. 81, 'Anonymous Note', n.d. [1816?].

[8] Musgrave, *Head Gardeners*, pp. 44–51.

had journeyed from the Scottish lowlands to England as a young man, and initially worked at Chiswick for the Earl of Burlington. James Lee probably trained with Philip Miller (1691–1771) at Chelsea Physic Garden,[9] and worked for the Duke of Argyll at Whitton Place, Twickenham and then for the Duke of Somerset at Syon House.[10] The varied training, in a botanical garden of high renown and in the gardens of influential and wealthy patrons, provided both partners with the experience and intellectual background necessary to run a successful nursery. It was this combination of skill and academic knowledge that ultimately allowed them to become pre-eminent in the British nursery trade.

An aptitude for learning was not enough on its own, however. Patronage was key to Kennedy and Lee's early success. Their aristocratic employers provided not just financial support but also connections to potential customers and other patrons. Kennedy's employers Robert and Dorothy Boyle, the Earl and Countess of Burlington, were important sponsors of horticulture; the countess was also Lady of the Bedchamber to Queen Caroline, whose zeal for gardening was widely celebrated. This provided a fillip to the commercial activities of businesses such as the Vineyard Nursery.[11] Further, wealthy patrons provided opportunities for education. James Lee is said to have had free use of Argyll's extensive library in London, where he would have had the opportunity to supplement his practical skills with studies in botany and architecture.[12] The different types of expertise that Kennedy and Lee accumulated through their work as gardeners, and through their education in the science and aesthetics of horticulture, contributed to the qualities of their partnership when they established the Vineyard Nursery in the mid-1740s. The skills they acquired were essential to the success and longevity of the nursery. Tracing the development and circulation of such skills is central to the argument of the book overall.

The day-to-day workings of the Vineyard Nursery are difficult to fathom – extant archival records regarding the nursery are very slight – but one key source does still survive. Kennedy and Lee, like many of their associates, regularly published catalogues listing the plants they had for sale. Dry in tone and plain in appearance, these little booklets, which

---

[9] Willson, 'Farming, market and nursery gardening', p. 92.

[10] Johnson, *History of English Gardening*, p. 216; Webber, *Early Horticulturalists*, pp. 91–92.

[11] Barrington, *On the Progress of Gardening* (1785), pp. 15–17; Colton, 'Merlin's Cave and Queen Caroline', pp. 1–20; Drayton, *Nature's Government*, pp. 38–39; Egerton, 'Boyle, Dorothy'; Kingsbury, 'Boyle, Richard'.

[12] Henrey, *Botanical and Horticultural Literature*, vol. 2, pp. 354, 658; Thornton, 'Sketch of the life of James Lee' (1810), p. vi. Argyll combined a career in politics with extensive scholarship. His house in London was known as 'the Library', and he was a Trustee of the British Museum from 1753 to 1761. Murdoch, 'Campbell, Archibald'.

were issued every few years, appear to say little about the culture of horticultural commerce. The 1774 *Catalogue*, for example, simply listed plant names in English, with no preface, illustrations, or other explanatory notes.[13] But the itemised plants nevertheless speak volumes about the customers that the Vineyard Nursery sought to attract, and thus give us a first impression of the direction in which horticultural commerce was orientated.

While the nursery's best-known patrons were mostly British aristocrats, Kennedy and Lee catered for a varied consumer group concentrated at the middling and upper end of the market. The nursery sold hardy trees and shrubs, herbaceous plants and bulbs, as well as a great range of exotic plants for hothouses. The *Catalogue* also listed a large number of edible plants for cultivation in kitchen gardens, and supplied seeds for trees and perennial and annual flowers. The final section was devoted to 'seeds to improve land', and concluded with a note advertising that they also sold 'Matts and all sorts of Garden Tools'. Several types of plants were sold 'in collections', which were pre-prepared mixtures of fruits or bulbous flowers, readily selected to please either the palate or the eye, as appropriate. In selling ornamental trees, shrubs and flowers, tender hothouse exotics, kitchen-garden plants and even seeds for agricultural use, the business could supply the full range of plants required for town gardens and the average landed estate.[14]

Located far from the worst of the coal-smoke, but still close enough to reach the city with ease, the Vineyard Nursery of Hammersmith was ideally situated to serve this wide range of consumers. Finding the right location for a nursery business was not easy: traders needed to be conveniently located for their customers, but also had to have ample space and protection from metropolitan pollution. As shown in Map 1.1, several of the upper-end nurseries solved this problem by operating from two locations, with a shop in the town and a garden further out in the countryside. But Kennedy and Lee found a different solution, keeping shop and garden together in the well-situated suburb of Hammersmith (#33 on Map 1.1). Set near the main turnpike road leading to the west of England, and close to the River Thames, the nursery was accessible both to Londoners and to those from villages and towns to the west of the city: the Vineyard was just downstream from Twickenham, for example, where there were several important aristocratic gardens.[15]

---

[13] Kennedy and Lee, *Catalogue* (1774).

[14] Laird, *Flowering*, pp. 8–13; Wilson and Mackley, *Creating Paradise*, p. 58.

[15] For examples, start with: Batey, *Alexander Pope*; Laird, *Flowering*, pp. 163–172; Mowl, *William Kent*, pp. 176–187.

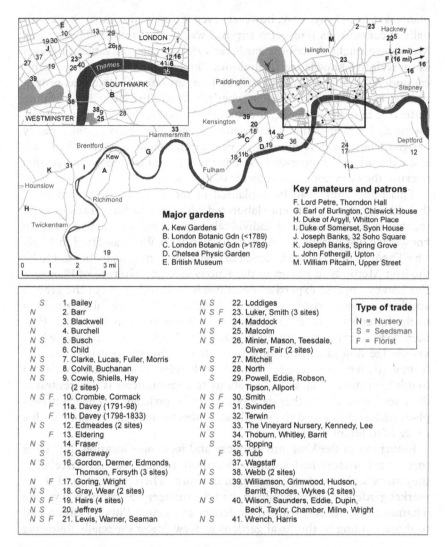

Major gardens

A. Kew Gardens
B. London Botanic Gdn (<1789)
C. London Botanic Gdn (>1789)
D. Chelsea Physic Garden
E. British Museum

Key amateurs and patrons

F. Lord Petre, Thorndon Hall
G. Earl of Burlington, Chiswick House
H. Duke of Argyll, Whitton Place
I. Duke of Somerset, Syon House
J. Joseph Banks, 32 Soho Square
K. Joseph Banks, Spring Grove
L. John Fothergill, Upton
M. William Pitcairn, Upper Street

S	1. Bailey
N	2. Barr
N	3. Blackwell
N	4. Burchell
N S	5. Busch
N	6. Child
N S	7. Clarke, Lucas, Fuller, Morris
N S	8. Colvill, Buchanan
N S	9. Cowie, Shiells, Hay (2 sites)
N S F	10. Crombie, Cormack
F	11a. Davey (1791-98)
F	11b. Davey (1798-1833)
N S	12. Edmeades (2 sites)
F	13. Eldering
N S	14. Fraser
S	15. Garraway
N S	16. Gordon, Dermer, Edmonds, Thomson, Forsyth (3 sites)
N F	17. Goring, Wright
N S	18. Gray, Wear (2 sites)
N S F	19. Hairs (4 sites)
N S	20. Jeffreys
N S F	21. Lewis, Warner, Seaman

N S	22. Loddiges
N S F	23. Luker, Smith (3 sites)
N F	24. Maddock
N S	25. Malcolm
N S	26. Minier, Mason, Teesdale, Oliver, Fair (2 sites)
S	27. Mitchell
N S	28. North
S	29. Powell, Eddie, Robson, Tipson, Allport
N S F	30. Smith
N S F	31. Swinden
N S	32. Terwin
N S	33. The Vineyard Nursery, Kennedy, Lee
N S	34. Thoburn, Whitley, Barrit
S	35. Topping
F	36. Tubb
N	37. Wagstaff
N S	38. Webb (2 sites)
N S	39. Williamson, Grimwood, Hudson, Barritt, Rhodes, Wykes (2 sites)
N S	40. Wilson, Saunders, Eddie, Dupin, Beck, Taylor, Chamber, Milne, Wright
N S	41. Wrench, Harris

Type of trade

N = Nursery
S = Seedsman
F = Florist

Map 1.1 London plant traders operating 1760–1800, with major gardens, key amateurs and patrons.

Nurseries like that run by Kennedy and Lee specialised in the cultivation and sale of exotic plants, many of which had been introduced only very recently to Britain. It was the task of plant traders to work out how best to cultivate specimens within their new climate and soil, eventually with a view to selling them on commercially. Naturalising plants was not always easy, however: even if a new introduction was initially successful, its long-term survival was not assured. Following Joel Mokyr, we might describe a gardener's skill in the cultivation of plants as prescriptive knowledge (techniques: 'knowledge how') and a botanical understanding of classificatory systems and nomenclature as propositional knowledge ('knowledge what').[16] While the latter is a defining feature of botanical science, the former should be recognised as equally essential, for propositional knowledge *of* and *about* plants is also partly derived from the development of prescriptive techniques for their care and cultivation.[17] Upper-end gardeners and plant traders were essential because they developed both forms of knowledge.

This relationship can be explained in terms of the development of 'hybrid expertise', a concept elaborated within the history of science and technology to describe how individuals might combine the two forms of knowledge just outlined.[18] As nurseryman John Webb asserted in 1760, 'the Theory joined to the Practice of Gardening, may justly be said to be the best Methods to obtain to the Knowledge and Culture of Plants.'[19] For botany, 'hybrid expertise' describes how knowing about the growth and living characteristics of a plant might contribute useful information to the botanical project of developing systems and classifications. Further, hybrid expertise was most likely to develop in situations where knowledge flowed in both directions: from scholars to gardeners, and from gardeners to scholars.[20] We might therefore add a third dimension to this hybridity: social skill. The ability to communicate is also central to the circulation and further development of knowledge.[21] The shrewdest plant traders developed the social astuteness to share information with a range of different social groups.

Returning to the Vineyard Nursery and its choice location, it is clear that its proprietors had ample opportunity to observe and share what they knew with a wide metropolitan network. There were around thirty market gardens in Hammersmith, and the nursery was connected via the Thames to botanically trained gardeners at Chelsea Physic Garden and to those tending to the royal gardens at Kew (most especially following

---

[16] Mokyr, *Gifts of Athena*, pp. 4–13.

[17] This is a slight inversion of Mokyr's formulation of the relationship between propositional and prescriptive knowledge. Mokyr argues that prescriptive knowledge 'has to have an epistemic base in Ω [propositional knowledge]'. This may be so for technological development (the subject of Mokyr's book), but the relationship between the two, as Mokyr himself acknowledges, is not completely clear-cut. For observational sciences, the relationship is surely the other way round. See: Mokyr, *Gifts of Athena*, p. 13.

[18] Klein and Spary, *Materials and Expertise*.   [19] Webb, *Catalogue* (1760), p. iii.

[20] Klein and Spary, 'Introduction'.   [21] Mokyr, *Gifts of Athena*, p. 7.

the creation of a botanic garden there in 1759).[22] In addition, Hammersmith had been home since 1751 to Lancelot 'Capability' Brown (c. 1716–1783), eighteenth-century Britain's pre-eminent landscape architect. Situated in such distinguished horticultural environs, the nursery could be sure of receiving a steady influx of new plants, information and clientele. The business run by Lewis Kennedy and James Lee was distinguished because these two socially astute nurserymen fused their practical experience, developed in the major gardens and nurseries in and around London, with botanical knowledge accumulated in the libraries and botanical collections of their patrons.

Andrieux and Vilmorin

Kennedy and Lee's Vineyard Nursery was one of a number of top-end horticultural businesses in London that jostled against one another for a share of the market. Paris, by contrast, saw the rise to prominence of one single company rather than multiple rival establishments. The nursery run by Jeanne-Marie-Adélaïde d'Andrieux (1756–1836) and her husband Philippe-Victoire Lévêque de Vilmorin (1746–1804) eclipsed local competition and ultimately enjoyed an outstanding longevity: the business remained within the family until the 1970s, and indeed still continues as a company, albeit under different management. Investigating its eighteenth-century origins reveals that the same triad of practical skill, botanical expertise and good social connections was equally as important in France, but that the way in which these factors were deployed varied. The plant trade developed more slowly in France compared to across the Channel. This meant that the earliest French commercial promoters of rare and exotic plants enjoyed tremendous opportunities for innovation.

Adélaïde d'Andrieux and Philippe-Victoire Lévêque de Vilmorin made substantial changes to a family business that was already well established by the time that they took over direction in the early 1770s. The Andrieux family had sold seeds and plants since the 1720s from a shop on the Quai de la Mégisserie in central Paris (the present-day Vilmorin company still has a shop on the same Quai). The business had gradually expanded and procured land on the Rue de Reuilly, Faubourg Saint Antoine, where they established a larger nursery garden. The two sites are indicated as #1 on Map 1.2.[23]

Adélaïde d'Andrieux's personal history is opaque: only a tiny amount of biographical information about her exists. We know that she was born

---

[22] Musgrave, *Head Gardeners*, p. 39.
[23] SV, Vilmorin – Vieux Documents, Annonce (1786).

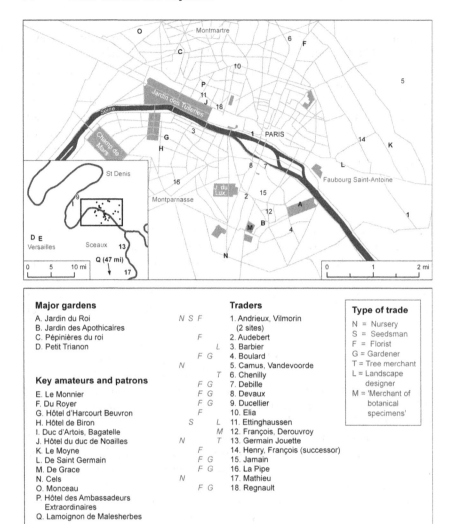

Map 1.2 Parisian plant traders operating 1760–1800, with major gardens, key amateurs and patrons.

in 1756 and was appointed 'maîtresse grainière' of the Parisian nursery company run by her father, Pierre d'Andrieux, in 1773. She vanished almost completely from the historical record after her marriage to Vilmorin in 1774; her mysterious disappearance opens up questions about the ways in which the horticultural trades were gendered, which will be discussed in the final section of this chapter. We know somewhat more, however, about her husband, Philippe-Victoire Lévêque de Vilmorin, who moved to Paris from the Lorraine in the 1750s or 60s.[24] Vilmorin was not originally a nurseryman: he trained as a doctor, and struck up a friendship with Pierre d'Andrieux at a botany course at the Jardin du Roi in Paris.[25] Marriage to a master's daughter was a common means of entry into trade for men who lacked formal instruction through the apprenticeship system.[26] Following Philippe-Victoire's death in 1804, the business was run by their son André Lévêque de Vilmorin (born 1776), quite possibly with help from his mother. Although I will use 'Vilmorin' as shorthand to refer to the nursery, it is important to underline that Adélaïde d'Andrieux may well have contributed to the family business until the 1820s or 1830s.

Andrieux and Vilmorin were innovators within French horticulture, playing a significant role in stimulating a new trade in exotic plants in France from the mid-1770s onwards.[27] They began from a good base, for the existing d'Andrieux business was already relatively high-ranking: Adélaïde's father Pierre had a royal *brevet* as seedsman and botanist to the king.[28] Yet when Adélaïde d'Andrieux became *maîtresse grainière*, the business mostly sold 'common' species.[29] Barely ten years later, Andrieux and Vilmorin could promise to supply a much wider range of rare plants, shrubs and trees.[30] Vilmorin is credited with being 'more innovative' than his rivals, having 'conceived of the idea of... commercialising species of plants valued in agriculture, but which still didn't exist [as garden plants] other than in the botanical gardens and in those of a small number of amateurs.'[31] Andrieux and Vilmorin were also very active in promoting the agricultural cultivation of European plants that were previously unknown in France, including the mangelwurzel (*Beta vulgaris*) and the frost-resistant Swedish swede (*Brassica napobrassica*).[32]

---

[24] We even have a description of him, from a passport issued in 1798: at fifty-two, Vilmorin had grey-brown hair, an oval face, a 'well-formed' nose and an 'ordinary' mouth and was 1 m 68 cm (5' 6") tall. AN, F/7/3580, Police Générale: Demandes des Passeportes J-Z 1793–1818, Vilmorin, 13 Vendémaire An 7 (4 October 1798).

[25] Heuzé, *Les Vilmorin*, pp. 5–8; Silvestre, 'Notice biographique', p. 194.

[26] Crowston, 'From school to workshop', p. 46.

[27] Silvestre, 'Notice biographique', p. 194.

[28] Andrieux, *Catalogue de toutes sortes de graines*, title page. The original French is: 'Maître & Marchand Grainier-Fleuriste, aussi Botaniste ordinaire du Roi'.

[29] Silvestre, 'Notice biographique', p. 194.

[30] Andrieux, *Catalogue de toutes sortes de graines* (1760).

[31] Silvestre, 'Notice biographique', p. 194.

[32] Ibid., pp. 195–196; Stephens, 'On the application of mangel wurzel' (1830), pp. 229–335.

Vilmorin's next and most significant innovation was to arrange for the regular importation of exotic plants directly to France. As with Kennedy and Lee, patronage was key to making this possible: Vilmorin was fortunate to have the support of the royal minister and amateur natural historian Chrétien-Guillaume de Lamoignon de Malesherbes (1721–1794). From 1779 onwards, the pair introduced barrowfulls of non-native vegetables and plants to France, including the tulip tree (*Liriodendron tulipifera*) and the American sweetgum (*Liquidambar styraciflua*).[33] As a result of such initiatives, the nursery helped to transform both French agriculture and France's gardens.[34] By 1831, Andrieux and Vilmorin's current and past successes were such that the British horticultural writer John Claudis Loudon declared that they were 'the first [of their kind] in the world'.[35] And at the end of the nineteenth century, Philippe-Victoire Lévêque de Vilmorin was fêted as 'the veritable creator of a scientific commerce of seeds' in France.[36]

The Plant Trade

But was Andrieux and Vilmorin's nursery *really* as distinctive as their nineteenth-century admirers claimed? And what was the contemporary horticultural world like, into which enterprises such as those run by Kennedy and Lee and Andrieux and Vilmorin introduced such innovative, 'scientific', practices? To understand the singularity of both these companies, we must delve into the earthy world of the plant trade, examining its history and geographical spread and the ways in which specimen supply networks were organised.

Agricultural and medicinal plants, so essential for human survival, have been swapped, sold and even stolen for thousands of years. Most eighteenth-century gardening books began with a history of the 'progress' of gardening since classical and biblical times; the authors conventionally mentioned the Garden of Eden and the Hanging Gardens of Babylon as noteworthy precedents to their own labours.[37] However, the motivations behind the creation of horticultural collections had changed

---

[33] Heuzé, *Les Vilmorin*, p. 12; Silvestre, 'Notice biographique', pp. 195–196.

[34] Heuzé, *Les Vilmorin*, p. 13. See also: MNHN, Ms. 318, Dossier XV, Vilmorin, Mémoire sur les semis d'arbres et d'arbustes (n.d., before 1792).

[35] John Claudis Loudon, *Gardener's Magazine* (1831), quoted in Heuzé, *Les Vilmorin*, p. 32.

[36] Between 1780 and 1899, Philippe-Victoire Lévêque de Vilmorin and his descendants were successively made members of the Société Nationale d'Agriculture de France, and were each awarded the prestigious Ordre National de la Légion d'Honneur. Heuzé, *Les Vilmorin*, pp. 12, 32.

[37] Anon., *Jardinier Portatif*, pp. 97–99; Lee, *Introduction to Botany* (1765), pp. iii, v–x; Weston, *Tracts* (1773), Introduction.

over time. Live plants had been gathered together since the Renaissance, either as medicinal collections or, more generally, as living outdoor counterparts to cabinets of curiosity. The seventeenth century saw collectors captivated by the so-called 'florists' flowers': bulbous plants that included hyacinths, auriculas and, most famously, tulips.[38] Florists' flowers were certainly still the subjects of fascination in the eighteenth century, and Chapter 3 will explore the culture surrounding florists' collections in more detail. In the eighteenth century, however, members of the middling and upper ranks largely shifted attention away from bulbous plants to exotic plants and trees. The 'new' exotics, which initially sold for exceptionally high prices, included specimens such as the beautiful *Acacia* and the otherworldly *Banksia*, wrenched from New World landscapes of the Americas and Australasia (from 1768), and the ancient *Ginkgo biloba*, uprooted from the hinterlands of China. Significantly, these non-native plants were assimilated within new cultural contexts, as well as new climatic ones. Examining how and why they became the subject of widespread consumption is key to comprehending the reception that such specimens received, and how they served the fortunes of the elite plant traders.

These plants were beautiful specimens, and thus held an intrinsic appeal in their own right. But they also possessed three characteristics important in stimulating consumer interest: novelty, variety and changeability.[39] The traders who are the subject of this book became adept at further developing these features and driving up the fashionability of each object. But, significantly, the plants were *also* perceived as specimens of scientific interest. Science itself, in eighteenth-century Europe, was becoming modish, and the possession of commodities associated with Enlightenment carried a certain cultural cachet.[40]

The consumption of new ornamental plants expanded significantly in Enlightenment Europe, in part due to a broadening public engagement with science and scholarly culture. Expensive scholarly collections were, and still are, primarily the preserve of wealthy individuals or institutions. Yet by mid-century, the middling ranks possessed the educational, cultural and economic capital with which to form their own botanical gardens or cabinet collections, albeit on a smaller scale.[41] The social history of the nursery trade is thus part of the history of the

---

[38] Conan, 'Histoire des jardins', pp. 970–972; Cunningham, 'The culture of gardens'; Goldgar, *Tulipmania*.

[39] For a comparable analysis of tulip consumption, see: Bianchi, 'In the name of the tulip', pp. 88–102.

[40] Easterby-Smith, 'Selling beautiful knowledge', pp. 531–543; Lynn, *Popular Science*; Stewart, *Rise of Public Science*.

[41] Lee, *Introduction to Botany* (1765), Preface.

uptake in consumption that characterised the later eighteenth century. Piecing together, from the limited archival information available, what nurseries existed and what the differences were between them is important for understanding how hybrid expertise emerged and for appreciating the unprecedented social opportunities that developed for relatively low-ranking individuals. Where and how widespread were the nurseries? Who were their proprietors, and in what ways were traders differentiated from one another?

The term 'plant nursery' ('*pépinière*' in French) does not necessarily refer to a commercial business. Most nurseries, indeed, were privately developed within specific gardens and farms and were primarily intended to produce specimens for individual use or very localised exchange. Selling plants and seeds on a wider scale was a secondary development. Commercial plant nurseries thus supplemented existing methods of obtaining and circulating seeds, and they had to offer something distinctive. In southern England, for example, where there were many large estates, commercial nurseries initially fulfilled the wishes of wealthier customers rather than the wants of everyday farmers or gardeners, who would only occasionally purchase seeds.[42] In France, we know of only one or two commercial nurseries near Paris, all from seventeenth-century records (although there were undoubtedly more).[43] In both countries, right through the eighteenth century, the arrangements for obtaining rare plants continued to be made by gardeners and stewards on an ad hoc basis.

The geographical distribution of the commercial nurseries differed between the two countries, and this had important implications for the development and distribution of expertise within each. Over the course of the long eighteenth century, the British nursery trade expanded outwards from London, Edinburgh and two provincial centres: York and Oxford. As transportation links improved, the nurseries ultimately formed a relatively well-integrated national network.[44] Records ·have been uncovered for sixty-five prominent nurseries between 1677 and 1800, and this corresponds to only a fraction of the total number of nurseries during the period.[45] By the latter half of the century, the new national network of nurseries could supply an exciting range of exotic

---

[42] Le Blanc, *Letters on the English and French Nations*, vol. 1, p. 321, quoted in Bourde, *Influence of England*, pp. 18–19; Weston, *Tracts*, pp. ii–iii; Young, *Travels in France*, pp. 18, 24 (May–June 1787).

[43] Bourde, *Agronomie et Agronomes*, vol. 1, p. 195; Heuzé, *Les Vilmorin*, pp. 3–4.

[44] Harvey, *Early Nurserymen*, ch. 4.

[45] Harvey, *Early Horticultural Catalogues*, pp. iii–iv. Harvey emphasises that he was unable to perform a fully exhaustive search.

plants to consumers across the country. Compared to France, this development represents an absolute boom.

The French plant trade, by contrast, emerged much more gradually and was strikingly localised.[46] French writers on horticulture and agriculture bemoaned their country's inferiority compared to its neighbours, focusing on the lack of diversity of the species cultivated in its gardens and fields.[47] In his *Essais d'Agriculture* (1778), Louis-François de Calonne lamented that 'most often, we limit ourselves to the species which are common in the climate where we live and on this point, our ideas are not carried any further.'[48] The same point was reiterated throughout the decades that followed, with writers repeatedly explaining that, with the exception 'of M. *Vilmorin*', France lacked well-educated, innovative nurserymen. Amateur plant collectors complained that they had to rely on suppliers from Holland and England, who (fortunately) sent 'enormous catalogues'.[49] In 1805, Georges Dumont de Courset explained that the British were horticulturally superior because

the subject of exotic vegetables, which the majority of people view as an insignificant object of trade, is become, [in Britain] ... a branch so lucrative that in a small amount of time they are able to make a profit of three to four hundred per cent. Such has always been the genius of this industrious nation.[50]

The few French gardeners and nursery companies that cultivated rare or exotic plants, 'in imitation of the English', were the exception rather than the rule.[51]

Other writers speculated as to why the existing trade in plants was less extensive in France than in Britain. Vilmorin's obituarist, Augustin-François de Silvestre, claimed that British nurseries boasted a greater profusion of exotic specimens because they had better access to

---

[46] Anon., *Jardinier Portatif* (1772), p. 19; Harvey, *Early Nurserymen*, p. 60.

[47] French commentators who remarked on the apparent superiority of British agriculture and horticulture include the Abbé Le Blanc, the Abbé Prévost and the Marquis de Gouffier, as well as the articles on agriculture in the *Encyclopédie*. See also: Bourde, *Influence of England*, ch. 2.

[48] Calonne, *Essais d'agriculture*, p. 96. Louis-François de Calonne (1714–1793) was an *Avocat* at the Parlement of Paris and *Grand-bailli* of Vitry-sur-Seine. He was a member of several academic societies, including the Royal Society of London.

[49] *Feuille du Cultivateur*, 30 Janvier 1793, pp. 38–40. The source refers specifically to 'Holland' and 'England' (rather than United Provinces and Britain).

[50] Dumont de Courset, *Le Botaniste Cultivateur* (1805), vol. 5, pp. 8–9. Despite these accolades, it is important to note that British gardeners also depended on the United Provinces for specimens. In the 1790s, Kennedy and Lee purchased several specimens of South African irises and gladioli that had been imported and bred by the Dutch nursery Voorhelm & Co. See: Andrews, *Botanist's Repository* (1797), pls 27, 38, 45. The Dutch nurseryman G. Eldering had his flowers sold in the London shop of a 'Mr Paas' in High Holborn. See: BM, Banks and Heal Collection of Printed Ephemera, Banks 74.9, Tradecard for 'G. Eldering' (1797).

[51] Calonne, *Essais d'agriculture*, p. 95.

suppliers. Britain's maritime trade, he noted, was more extensive than that of the French, and this therefore furnished the nurseries with a greater quantity of contacts overseas.[52] It was certainly the case that, by November 1805, when Silvestre presented his tribute to Vilmorin's life, French maritime power had been decisively eclipsed by that of the British. However, Silvestre's assumption that the exotic plant trade was determined by supply is not borne out by the evidence. During most of the eighteenth century, France's maritime trade was sufficiently far-reaching to have permitted nurseries to build up their own stock of exotic plants, had they wished to do so. France's colonial trade increased ten-fold between 1716 and 1787, and the quantity of plants and other natural history objects sent back to the country increased notably once France had stabilised control over her colonies, especially following the Peace of Paris in 1763.[53] The country possessed colonies and trading posts in parts of the world that sported great floral wealth, particularly in the Caribbean, along India's Coromandel Coast and in North America.[54]

French botanists were also very active in obtaining plants from across the globe; they sent out plant hunters and encouraged travellers to bring back interesting specimens.[55] The correspondence and account-books of the Jardin du Roi attest to regular arrivals of new plants from abroad, which were often shared between the Jardin and other royal gardens, as well as with a select number of elite private collectors such as Malesherbes.[56] Botanical gardens were created or expanded across the country, especially in port towns such as Nantes, Brest and Bordeaux, to house these new specimens as they arrived.[57] French nurseries certainly could have obtained more exotic plants directly from overseas if consumers had wanted them.

The lack of national interest in the cultivation of foreign species was both cause and consequence of the dearth of French commercial nurseries. Cultivating rare plants was expensive; developing a desire to do so was as much a question of internal supply as it was of demand. The limited distribution of nurseries, poor communications and internal customs barriers meant that it was difficult to transfer plants from

---

[52] Silvestre, 'Notice biographique', p. 195.

[53] Butel, *L'économie française*, pp. 81–83; Crouzet, *Britain Ascendant*, p. 296; Spary, *Utopia's Garden*, pp. 122–123.

[54] Butel, *L'économie française*, pp. 83, 87–88.

[55] McClellan and Regourd, *Colonial Machine*; Mukerji, 'Dominion, demonstration, and domination', pp. 19–33.

[56] For examples, see: AN, AJ/15/149, Dépenses pour le Jardin, 1760 à 1793; AN, AJ/15/511, Envois de Graines, plantes, minéraux etc, de pays étrangers au jardin du roi; AN, 399/AP/97, Lettres de Malesherbes, de l'abbé Marc, de Saint-Jean de Crevecœur et du duc d'Harcourt 1782.

[57] Bourguet and Bonneuil, *De l'inventaire du monde*, p. 21.

one part of France to another. With the exception of the port towns, few provincial nurseries stocked large collections of healthy exotic specimens – and so the wealthy consumers who actually sought out new exotics often obtained them from abroad.[58]

Although new plants arrived regularly in France from the 1760s onwards, the national network of commercial nurseries remained in its infancy until the early nineteenth century. During the Old Regime, only a small handful of provincial nurseries had the means to serve customers beyond their immediate locality.[59] Some scientific societies and local governments formed initiatives to set up nurseries in order to introduce or improve certain plants. For example, the Estates of Languedoc established a network of nurseries in 1723 to improve mulberry cultivation in the province, and in the 1760s the Société d'Agriculture created a nursery at Sens with the intention of encouraging the cultivation of a wider range of trees.[60] These initiatives, however, were focused on agricultural development rather than horticulture, and did not improve gardeners' access to new plants at a national level.[61] The connections between these places and the capital city were loose, to say the least. It was not until the Revolution that the Paris-based botanists systematically sought to find out exactly *what* was cultivated or preserved in the provincial gardens.[62] From an international perspective, the capital often eclipsed these provincial places of learning, and most British travellers and scholars focused their energies on travelling to, and obtaining contacts within, Paris.[63] It is France's capital, then, which primarily forms the focus for the French part of this study.

Metropolitan Nurseries

The relative importance of the Parisian nurseries to French botany is underlined by the correspondence records kept by André Thouin (1746–1824), the head gardener at the Jardin du Roi. Thouin carefully itemised

---

[58] Anon., *Jardinier Portatif* (1772), p. 19; Silvestre, 'Notice biographique', p. 194.

[59] The slow development of a national nursery network is discussed in: AN, F/10/371, M. Féburier, Observations sur les pépinières du Gouvernement, 2 January 1808.

[60] Anon. [De Grace?], *Bon Jardinier* (1768), pp. 189–190; Livesey, 'Botany', p. 70; Weston, *Tracts*, p. iv.

[61] AN, F/10/371, M. Féburier, Observations sur les pépinières du Gouvernement, 2 January 1808.

[62] Silvestre, 'Notice biographique', pp. 201–204.

[63] This is not to say that French provincial centres did not attract international attention – botanists in Montpellier, as James Livesey has shown, circumvented Paris to form their own links beyond France's borders. But Paris nevertheless remained the focus of most foreign interest. On Paris as a capital of Enlightenment, start with: Romano and Van Damme, 'Sciences et villes-mondes', pp. 7–18; Van Damme, *Paris*. For Montpellier, see: Livesey, 'Botany'.

all the letters that he received from, and sent to, named correspondents, and his records show that by the 1770s and 80s he corresponded with at least twenty commercial nurseries in Paris.[64] Most of these are indicated on Map 1.2, and most were concentrated at the lower end of the market. The average number of letters exchanged with each is only three per business, which suggests that they may have possessed little of interest to the Jardin. Anecdotal evidence underscores the impression that few had a good reputation. The Scottish immigrant gardener Thomas Blaikie (1750/1–1838) was shocked to discover in Paris that his aristocratic employers obtained plants from unscrupulous characters such as the 'jardinier fleuriste du Dauphin' M. Henry, who, despite his royal *brevet*, apparently spent most of his time 'at his bottle' in the local tavern rather than in his nursery on the Rue de la Roquette (#14 on Map 1.2).[65] Blaikie eschewed Henry and other such suppliers, creating private (non-commercial) nurseries in his patrons' gardens. Like the collectors mentioned previously, he also obtained plants from plant traders overseas – most especially from Lewis Kennedy and James Lee.[66]

Most French gardeners seeking a wider range of plants used horticultural manuals, especially the *Almanach du Bon Jardinier*, to find out the names of more reliable suppliers. The *Bon Jardinier*, which was (and still is) one of the best-known gardening manuals in France, printed annual lists of plants, offered brief instructions about their cultivation and, crucially, gave the names and addresses of collectors and nurserymen in and around the capital.[67] The people it cited most frequently included the apothecary M. Descemet, who cultivated an extensive collection of exotics in the Jardin des Apothicaires on the rue de l'Arbalète (B on Map 1.2), a M. Mathieu at Villeneuve-le-Roi (#17), Germain Jouette, who specialised in growing exotic trees in his nursery at Vitry-sur-Seine (#13), and Andrieux and Vilmorin (#1).[68]

Indeed, the superiority of the latter two nurseries is clear both in the *Bon Jardinier* and in Thouin's correspondence records. Germain Jouette's nursery was regularly promoted in the earlier editions of the *Bon Jardinier*, and his catalogue was reprinted as an appendix in the 1760s and 70s. Although Pierre d'Andrieux apparently contributed to the

---

[64] MNHN, Ms. 314, État de la correspondence d'André Thouin.

[65] Blaikie, *Diary*, 13 January 1778, p. 143.

[66] Blaikie, *Diary*, p. 173 (nursery); Livesey, 'Botany', p. 70; Traversat, 'Les pépinières', p. 141.

[67] The *Bon Jardinier* was first published in 1755 and (apart from a small hiatus during the Revolution) has been published annually up to the present day. Grand-Carteret, *Les Almanachs Français*, p. 66.

[68] Anon. [De Grace?], *Bon Jardinier* (1768), p. 167 (Descemet), pp. 181–188 (Germain Jouette); Anon., *Le Bon Jardinier*, p. 192 (Mathieu).

almanac between 1755 and 1778, the name 'Andrieux and Vilmorin' was not printed until 1785. But this date marks the moment when the nursery seems to have decisively eclipsed its metropolitan rivals: its name and catalogue were exclusively featured in the *Bon Jardinier* at the expense of all others.[69] Thouin also corresponded extensively with both nurseries. Germain Jouette wrote seventy-four letters to Thouin between 1766 and 1687, and Andrieux and Vilmorin wrote ninety-one letters between 1767 and 1791, few of which, unfortunately, have survived.[70]

By contrast, the late eighteenth century was an exciting time to be a plantsman in Britain's capital. Aided both by rising national and international demand and the influx of novel supplies, many of London's nurseries gained distinction for the rarity and range of the plants they advertised, and for the expertise of their proprietors. From the surviving catalogues and trade cards, we know of eleven nursery or seeds businesses that advertised in London between 1700 and 1740. The number of advertising nurseries nearly quadrupled in the 1760–1800 period to forty-one, eight of which had existed prior to 1740.[71] As indicated by the multiple surnames listed against single sites on Map 1.1, a single nursery garden was often run in a partnership between different owners. Partners would usually work together for a time, before the direction of a garden was handed over to the new partner as a successor. Alternatively, nurseries were passed down within families. The number of nurseries in the city increased partly as a result of the growth of population and prosperity within London: greater demand for flora from city-based consumers encouraged more entrepreneurial gardeners to set up shop. And as the number of nurseries increased, so did the level of expertise and the opportunities for developing further contacts with other gardeners and traders. Around ten or eleven of the nurseries, indicated in bold type on Map 1.1, could be considered to be 'upper end'.

Britain also saw rapid standardisation in the prices charged for plants as its national nursery network expanded and consolidated. In the 1790s, herbaceous plants cost between three and six pence; shrubs were one or

---

[69] Anon. [De Grace?], *Bon Jardinier* (1768), pp. 180–181; de Grace, *Bon Jardinier* (1785). A lengthy advertisement for the nursery on pp. 350–351 explained that '*Vilmorin-Andrieux*...is the only [nursery] in a state to supply everything that is the most rare...either from France or from foreign Countries'.

[70] The number of letters sent did not seem to depend on geographical proximity: Andrieux and Vilmorin lived nearer to Thouin than many of their rivals, yet they sent him more letters. Their garden was just across the river from the Jardin du Roi, and their boutique was about thirty minutes' walk away. MNHN, Ms. 314, État de la correspondence d'André Thouin.

[71] Note that these figures only reflect those businesses whose advertisements have survived, or which were mentioned in other contemporary sources. Given the paucity of sources it is likely for both periods that the actual number of nurseries was much higher.

two shillings.[72] For the traders who had access to the best and rarest specimens, however, selling plants could be very lucrative indeed. Nurseryman Daniel Robertson of London advertised seeds of 'the Royal Balsam' in around 1790, which he sold 'in Papers of Five Shillings each', and which he described as 'a curious [plant] . . ., which far surpasses any Thing of the Kind ever seen in this Kingdom'.[73] James Lee asked for 10s. 6d. for the *Alstromeria pelegrina* in 1775, and prices could escalate even further. In the early 1790s, Lee was apparently selling his first fuchsias for a guinea each.[74]

In both Britain and France, the traders' canny formation of a conjoined expertise in horticulture and botany permitted the crafting of companies that were substantially different from those within the wider plant trade. In geographical terms, the distribution of this skill was somewhat different between the two countries. Britain saw the development of a national network of nurseries and gardens within which new information could be transferred and techniques further developed. Travel within this network was common, particularly from north to south, as gardeners searched for education and employment.[75] London and its environs grew in significance as a centre of expertise: writers on botany and horticulture constantly referred to practices observed among gardeners in the capital.[76]

Eighteenth-century France did not possess such a national network; here we are presented with a story of concentration of expertise. The skills developed within selected private gardens and commercial nurseries in Paris were enhanced by formal botanical instruction at institutions such as the Jardin du Roi, which was, in fact, how Pierre

---

[72] Harvey, 'Commentary' in Galpine, *The Georgian Garden*, pp. 45–48.

[73] Bodleian, John Johnson Catalogue of Printed Ephemera, Trade Cards 14 (4), trade card for Daniel Robertson, gardener and nursery-man (n.d. [1790?]).

[74] James Lee told Joseph Banks that other nurserymen charged more for the *Alstromeria*, but that his was 'the faire [sic] price'. Sir Joseph Banks Archive Project, Nottingham Trent University, James Lee (Vineyard Nursery) to Joseph Banks, 5 August 1775 [Original in Yale University Mss.]. I am grateful to Dr Neil Chambers for sending this to me. *Fuchsia coccinea* was first described by William Aiton in the *Hortus Kewensis* (1789); in 1792, William Curtis noted in his *Botanical Magazine* that specimens could be purchased from the Vineyard Nursery. There are competing stories about how Lee came by his first fuchsia specimen. See: Willson, *James Lee*, pp. 28–30.

[75] Indeed, the migration was so intensive that a xenophobic debate blew up in the English press in the early 1780s, provoked by an alarmist article by 'Investigator' in *St James's Chronicle* no. 3363, which depicted London overwhelmed by an 'extraordinary inundation of Scottish gardeners'. The Scottish gardeners were defended by 'P. B. C.' in 'Inundation of Scottish Gardeners accounted for', *Gentleman's Magazine* (1783), vol. 53, p. 322.

[76] For examples, see: Abercrombie, *Complete Kitchen Gardener* (1789), p. 463; Andrews, *Botanist's Repository* (1797); Loddiges and Sons, *Botanical Cabinet* (1817).

d'Andrieux and Philippe-Victoire Lévêque de Vilmorin first met. By the late eighteenth century, Paris and its surrounding area were recognised as centres of horticultural and botanical proficiency.

Four significant features defined the eighteenth century as a special period for commercial horticulture in general, and for the elite nurseries in particular: the growing culture of consumption, the development of domestic trade, the concentration of expertise in capital cities and the increase in international mercantile connections. Coupled with the expansion of scholarly networks, these factors stimulated significant developments within the domestic plant trade in both Britain and France. The more elite British and French nurseries capitalised on this, directing consumer behaviours and deploying the new networks as means of developing and distributing skill and expertise. Their own formulations of hybrid expertise were, again, central: these innovative companies combined practical skills with a more abstract understanding of the science of botany. Perhaps even more importantly, as we will see later, they broadcast their proficiency among the wider public.

People in the Plant Trade

The accumulation of skills in both horticulture and botany, and the blending of these into hybrid expertise, was clearly crucial to the formation of an elite cohort of gardeners. But what sorts of people were able to join this select grouping? Without exception, the merchants presented themselves in their advertisements and correspondence as modest and unassuming. There was no mention of them being 'elite', although they did assert some superiority over their less-well-educated counterparts. Horticultural skill was also depicted as a male attribute. But such self-presentations paper over a more complex reality. Despite the conventional representation of the nursery trade as a male profession, wives and daughters did contribute alongside husbands and sons, albeit usually behind high garden walls. The 'humble' art of gardening was also deeply divided by socio-professional hierarchies and guild structures.

The frontispiece to the eighteenth edition of Thomas Mawe and John Abercrombie's *Every Man His Own Gardener* (1805) depicts British gardener John Abercrombie (1726–1806), standing proudly in a walled kitchen garden (Figure 1.2). Abercrombie rests one hand genteelly on a spade; with the other he gestures towards the garden. Behind him, a boy is hard at work hoeing, while a youthful couple – possibly the proprietors of the house and garden – promenade along the path. The social diversity within the gardening professions is evident in the juxtaposition of the two gardeners. The boy is a lowly labourer: he is a tiny figure in

JOHN ABERCROMBIE, Ætat 72.

Publish'd by C.G. & J.Robinson, as the Act directs May 1, 1800.

Figure 1.2 John Abercrombie (1726–1806), from Thomas Mawe and John Abercrombie, *Every Man His Own Gardener* (18th edn, London, 1805), frontispiece. Courtesy of the University of St Andrews Library = s SB453-M2E05.

the background relegated to manual work. Abercrombie, on the other hand, stands proud on the path and dominates the picture. The only visible distinction between this gardener and his employers is that he holds a spade; his clothes and demeanour associate him more closely with the elegant young couple than with the boy who grubs away removing weeds.

John Abercrombie was a head gardener who had been, he stated, 'sixty years a practical gardener'.[77] His father had been a market gardener in Edinburgh, and Abercrombie had accrued his own expertise – and evidently his wealth – through several decades of hard physical work coupled with an extensive education in horticulture and botany.[78] His background was thus not dissimilar to those of Kennedy and Lee. Abercrombie and his associates were the cream of their profession, and some even commanded relatively high social and civic positions: they counted the gentry, freemen, municipal officers and even mayors among their number.[79] They achieved respect for their skills as practical gardeners, especially with regards to the naturalisation of non-native plants, and for their intellectual command of cutting-edge botanical knowledge: in addition to James Lee's *Introduction to Botany* (1760), other nurserymen and head gardeners who published on botany included William Aiton, *Hortus Kewensis* (1789) and Conrad Loddiges, *Botanical Cabinet* (1817). Many more nurserymen and gardeners, such as James Gordon, were 'too modest' to publish, but were nevertheless respected as authorities on botany.[80]

These nurserymen and gardeners were not 'gentlemen', but they could claim membership of the 'genteel' classes. They possessed good educational levels, although this was largely achieved through their own initiative, and they had a greater capacity for financial investment than the retailers and craftspeople who made up the 'common trades'. In the 1760s, a head gardener in a provincial garden might earn around 1s. per day, but wages were substantially higher in London: James Lee paid his apprentices 8–12s. per week. In France, a gardener working at the Jardin du Roi in the late 1780s would earn around 9 or 10 *livres* a week. The cost of setting up as a master gardener in Britain, between £100 and £500, was roughly the same as that for glaziers and joiners.[81] The

[77] Mawe and Abercrombie, *Every Man His Own Gardener* (1805), title page.
[78] Johnson, *History of English Gardening*, p. 219; Longstaffe-Gowan, *London Town Garden*, pp. 157–158; Musgrave, *Head Gardeners*, pp. 48–51, 58–61.
[79] Nurseryman Francis Noble, for example, was Mayor of Newark in 1739 and 1752. Harvey, *Early Nurserymen*, p. 68.
[80] Aiton, *Hortus Kewensis*; Lee, *Introduction to Botany* (1760); Loddiges and Sons, *Botanical Cabinet*. On modesty, see: Grieve, *Transatlantic Gardening Friendship*, p. 22.
[81] Longstaffe-Gowan, *London Town Garden*, pp. 157–158; Musgrave, *Head Gardeners*, p. 41. At the top of the profession, the landscape gardener Richard Woods charged

'elite' nurserymen and gardeners were situated much more closely to other 'polite' sections of society, such as the professional classes and the lesser-landed gentry.[82]

They were also distinguished from the seedsmen, florists, nurserymen and gardeners who were not well versed in botany, and who cultivated a more mundane range of plants.[83] Writing in the 1760s, British merchant Peter Collinson (1694–1768) recalled with some frustration that 'after I had supplied the several persons [with American plants] . . . the next thing was "Pray sir, how and in what manner must I sow them . . . my gardener is a very ignorant fellow".'[84] Ten years later, Horace Walpole wrote in similar terms about his own gardener, who, despite having 'lived with me above five and twenty years . . . is incredibly ignorant and a mule'; he concluded by angrily declaring that 'the serpent . . . has reduced my little Eden to be as nasty and barren as the Highlands'.[85] In contrast to such 'ignorant fellow[s]', the 'better sort' of gardeners and plant traders discussed here were well educated in botany. In an earlier letter, Walpole wittily described how he had 'made great progress [in "planting"], and talk very learnedly with the nurserymen, except that now and then a lettuce run to seed overturns all my botany'.[86] Having achieved a relatively high level of social respectability, these learned nurserymen were on a par with the likes of John Abercrombie.

The social composition of the gardening professions in France was similarly varied. Even in the 1760s, the Andrieux nursery had already clearly distinguished itself from rival traders such as the drunkard M. Henry, described before. The full title of Pierre d'Andrieux's royal *brevet* was 'Master Seed and Flower Merchant, also King's Ordinary Botanist'; Philippe-Victoire Lévêque de Vilmorin used the same title,

1 guinea per day for his services in 1769. Laird, *Flowering*, pp. 21–22. For the wages paid to gardeners at the Paris Jardin du Roi, see: MNHN, Ms. 47 Joseph Martin, Copy of letter from André Thouin to the Comte de La Luzerne, 14 February 1788, f. 1v.

[82] Vickery, *Gentleman's Daughter*, p. 28.

[83] Unskilled garden labourers earned a few pence a day. At Wolterton in 1738, men earned 1s. 2d. in the garden (2d. more than builders' labourers) and women 7d. (1d. more than the women cleaning the newly constructed house). Wilson and Mackley, *Creating Paradise*, pp. 176, 172.

[84] Peter Collinson, manuscript reminiscences on his life, dated 1766, quoted in Swem, 'Brothers of the spade', p. 21. Collinson was a cloth merchant and amateur botanist who was made a Fellow of the Royal Society in 1728. His garden at Mill Hill (London) became renowned for the range and rarity of its content, thanks to his connections with the North American plant merchant John Bartam (discussed in Chapter 4).

[85] Walpole's gardener, like so many others in London, was Scottish. Horace Walpole to Lord Harcourt, 18 October 1777, from Toynbee, *Strawberry Hill Accounts*, note 5, p. 38.

[86] Horace Walpole to Henry Seymour Conway, 29 August 1748, in Toynbee, *Strawberry Hill Accounts*, note 5, p. 36.

adding '*pépiniériste*' (nurseryman) in 1778.[87] Vilmorin was not from a particularly elevated background: his father, who died in 1759, was apparently a farmer (the type is unclear).[88] Nevertheless, around the time of his entry into the Andrieux nursery business, Vilmorin saw fit to adopt a noble 'de' in his name. Self-ennoblement in this manner was relatively common among the professional classes in France towards the end of the Old Regime, and is indicative of the status that Vilmorin, as a nurseryman with a royal commission, felt permitted to claim.

British and French social structures, educational opportunities and attitudes towards entrepreneurial initiatives were not identical, but the social composition of the horticultural trades was nevertheless very similar. Both countries saw an increasingly marked division emerge between the 'common' gardeners and plant traders and those who distinguished themselves as 'elite'. The latter based their claims to a more elevated status on their expertise as plantsmen (both practical and intellectual), and they presented this in public as one aspect of a polite, genteel persona.

Women's Participation in the Plant Trade

The question of whether, and what, women contributed to the horticultural trades is relatively unclear, and perhaps all the more so for the more 'genteel' plant traders. In several ways, this more elevated branch of the profession seems to have barred respectable women from participating: gardening itself, of course, requires some heavy labour, which in the eighteenth century was construed as both demeaning and defeminising for a respectable woman. And although, as indicated by the picture of John Abercrombie, the more sophisticated gardeners and plant traders were primarily involved in management rather than physical work, the additional tasks they took on were no more favourable to female involvement.

Andrieux and Vilmorin tantalisingly present a picture of female direction right from the start. The nursery was founded by Adélaïde d'Andrieux's grandmother, Jeanne Diffetot.[89] It then passed down the female line, eventually to Adélaïde d'Andrieux herself. But Andrieux disappears almost entirely from the nursery's few surviving archival records

---

[87] Andrieux, *Catalogue de toutes sortes de graines* (1760) title page; Andrieux and Vilmorin, *Catalogue* (1778), title page. Andrieux and Vilmorin described themselves as 'Marchands Grainiers-Fleuristes & Botanistes du Roi, & Pépiniéristes'.

[88] Vilmorin's medical and botanical education in Paris may have been thanks to the patronage of his godfather Philippe Dessoffy de Cserneck, an army captain, who was presumably responsible for sending young Philippe-Victoire to Paris following his father's death in 1759. Heuzé, *Les Vilmorin*, pp. 6–7.

[89] Heuzé, *Les Vilmorin*, p. 6.

after her marriage in 1774. Her obscurity is underlined by the confusion that exists about even her lifespan. In his 1899 history of the Vilmorin family, biographer Gustav Heuzé incorrectly claimed that Andrieux died in 1780, and this has been repeated subsequently.[90] But Andrieux's gravestone, in the Vilmorin family tomb in the Père Lachaise cemetery in Paris gives her date of death as 1836; a date which is further corroborated by a post-mortem inventory of her possessions from 11–12 March of that year.[91] Nevertheless, it was Philippe-Victoire, not Adélaïde, who signed the surviving correspondence with customers and associates, and whose name was published beside articles on gardening in books and trade catalogues. And thus it is Vilmorin whose name has been remembered for posterity. Why, then, did an apparently capable master seedswoman become so completely removed from the family business?

We know very little about the events that punctuated the personal lives of Andrieux and Vilmorin. Adélaïde d'Andrieux may, for example, have become incapacitated by illness or accident, and thus unable to work. Yet the fact that she lived until she was eighty suggests otherwise. There are two other compelling reasons for why such a woman might disappear from the nursery's records. One relates generally to the gendered nature of work in eighteenth-century France (and Britain), and the other to the specific tasks that the proprietors of an elite nursery carried out.

Many of the skilled trades in France were founded on 'familial models' of participation. These, as we will see, paralleled but were not identical to the gendered nature of work in Britain. Most occupational groups in France were formally fixed and monitored by the guild system, and membership of several trades was restricted by gender.[92] Female family members might contribute illicitly to a trade or business, but as this labour formed a shadow economy, they were excluded from official records.[93]

There was no specific guild for nursery gardeners (pépiniéristes), and the men and women who made a living from the plant trade described themselves variously as seed-sellers, gardeners or market gardeners.[94]

---

[90] Ibid.

[91] AN, RE/LXXV/22 and ET/LXXV/1132, Inventaires après décès d'Adélaïde d'Andrieux, 11–12 March 1836.

[92] Goody, *The Culture of Flowers*, pp. 216–217; Kaplan, 'Social classification', pp. 177–185, 220–225 and *La fin des corporations*, pp. xiv–xv, 7, 47–49.

[93] Hafter, 'Women who wove', p. 45. See also: Locklin, *Women's Work* for a detailed discussion of the gendering of work (and especially of women as autonomous traders) in provincial France.

[94] Gardeners (*jardiniers*), market gardeners (*maraîchers*), seed-sellers (*grainiers* or *grainetiers*), florists (*fleuristes*) and bouquet-makers (*bouquetières*) each had their own

Adélaïde d'Andrieux legitimately gained the title 'maîtresse grainière' in 1773.[95] Given her expertise, and the fact that the company retained the name 'Andrieux' until the nineteenth century, it seems probable that Adélaïde did continue to contribute in some way even after her marriage. However, she lived and worked within a culture in which it was rarely appropriate to broadcast female labour overtly, and the nature of her contribution is thus obscure. Reasons of propriety are the mostly likely explanation for why Andrieux was removed from the public gaze following her marriage to Vilmorin in 1774.

The situation was only slightly different in Britain. As in France, women feature only occasionally in nursery records: I have not uncovered any surviving letters or bills sent to customers by women, and the term 'nurserywoman' was not used, as far as I am aware. The absence of nurserywomen from the archive should not, however, be taken at face value. Family connections were equally important in late eighteenth-century Britain as in France, and it was usual for women to contribute to the business. Relations between husbands and wives were often 'co-dependent' rather than patriarchal, but mercantile activity was represented externally by the head of the household, who was normally male.[96]

It is likely, then, that women did work in nurseries. Lewis Kennedy and James Lee were both married and lived with their families at the Vineyard Nursery. Lewis Kennedy married twice, first to Margaret Garioch (1756), then (following her death) to Margaret Aldrich (1773). He had seven children, including three sets of twins. James Lee and his wife Martha (1710–1779) had three daughters: Susannah (born 1748), Ann (1753–1790) and Mary (date of birth not known), and a son, James (1754–1824).[97] Although for the most part records do not survive detailing what (in addition to bringing up the numerous offspring) the female contribution may have been to the business, tasks

guild. In the late seventeenth century, the terms *pépiniériste* and *jardinier-pépiniériste* described a specialist who cultivated trees, yet by the 1760s they were commonly used in ways which roughly translate as the English 'nurseryman' (or woman). The connotation of expertise, however, was retained. Franklin, *Dictionnaire historique*, pp. 368, 500; Rey, *Dictionnaire historique*, vol. 2; Traversat, 'Les pépinières', p. 173.

[95] The word 'grainier' was used to distinguish seed-sellers from salt-sellers, who were also called 'grainetiers'. It has been replaced by 'grainetier' in modern French. Distinctions were made between at least four types of seed-seller: 'marchand des graines', 'épicier-grainetier', 'négociant en blé' and 'commissionnaires en graines et farines'. AP, D.5B6 1–3000 'Commerçants faillis'; Cayla, *Histoire des arts*, pp. 130–132; Franklin, *Dictionnaire historique*, p. 368.

[96] Barker, *Business of Women*, pp. 6–7.   [97] Willson, *James Lee*, pp. 6, 56.

such as account-keeping were widely considered suitable female employments. Light gardening was recommended to middle-ranking women as a healthful activity.[98]

The involvement of British women in running the upper-end nurseries, and their potential contributions to the distinctive blend of commerce and science cultivated within those nurseries, was encouraged by other factors. Dissenting religious culture was tremendously important. Almost all the elite nurseries in eighteenth-century London were run by Quakers, who (like other Dissenters) were notably progressive with regards to female education.[99] At least one of Lee's daughters, Ann, was trained in botany and studied botanical painting with Sydney Parkinson (c. 1745–1771), another Quaker who later sailed on Captain Cook's *Endeavour* voyage 1768–71.[100] Ann's depictions preserved and promoted the fragile, unusual plants that were cultivated in the Vineyard Nursery, and as such her work was essential for reinforcing the nursery's claims to be a scholarly institution. Several of her father's botanist correspondents, including Joseph Banks, Charles-Louis L'Héritier de Brutelle and Carl Linnaeus, requested her paintings of plants.[101] Kennedy and Lee's other children may well have shared a similar educational upbringing to Ann Lee, and the girls could have, like her, contributed to the scientific dimension of their trade in ways considered appropriate to their sex. The term 'nursery*man*' conceals the unrecorded contributions of wives and daughters, which were also central to sustaining the family business.

The people who devoted their lives to the commercial collection and cultivation of plants spanned a broad social gamut, from jobbing seed-sellers to well-educated gardeners and traders. The latter group achieved success both as traders and as participants in Enlightenment scholarly circles as a result of three main factors. Starting with the most significant, they combined their practical training as gardeners with an understanding of botany in order to develop their own form of hybrid horticultural expertise. For the elite traders, hybrid expertise consisted of practical

---

[98] For account-keeping, see: Davidoff and Hall, *Family Fortunes*, pp. 383–384. For gardening, see: CRO, D/HC/1/81, Catherine Mary Howard, 'Reminiscences for my Children' (1831), f. 19; Wakefield, *Reflections* (1798), pp. 91–92, 137–138. By contrast, the wives of full-time gardeners often sought employment at the same house as indoor servants. Longstaffe-Gowan, *London Town Garden*, p. 156.

[99] Mack, 'Religion, feminism', pp. 434–459.   [100] Allen, 'Parkinson, Sydney'.

[101] BL, Add. Ms. 8097, f. 192, L'Héritier de Brutelle (Paris) to Joseph Banks, 28 May 1789; LS, L4741, James Lee (London) to Carl Linnaeus, 23 October 1772; LS, L5155, James Lee (London) to Carl Linnaeus, 7 October 1775; LS, L5238, James Lee (London) to Carl Linnaeus, 4 October 1776.

and intellectual knowledge (expressed through their work in naturalising and cultivating rare plants) and social knowledge (expressed through their ability to communicate with diverse groups in the marketplace, and within more refined scholarly circles). Second, the metropolitan locations of their nurseries further facilitated access to new specimens, to expanding markets for these specimens and to local knowledge networks that ranged from the botanical to the practical. Finally, their success, both as traders and as social intermediaries, was the result of the intellectual, economic and practical contributions made by whole families. The story does not end here: the nurseries were also skilled at deploying their contacts within these networks to gain access to new specimens, and at stimulating and manipulating consumer demand. This was as much a question of commanding cultural knowledge as it was of skilfully handling the social and botanical knowledge described in this chapter.

2   Science, Commerce and Culture

In 1792, the Scottish nursery company Archibald Dickson and Sons published a catalogue of 'Hot-House, Green-House and Hardy Plants'. It opened its three-page 'Preface' with the following epigraph:

> Then spring the living herbs, profusely wild,
> O'er all the deep-green earth, beyond the power
> Of BOTANIST to number up their tribes[1]

Dickson's catalogue marked out new ground in the genre of horticultural publications, for it opened with a literary quotation. It also departed from the conventional structural divisions used in other commercial catalogues. Whereas lists of plants were normally subdivided according to different types of garden, most typically the shrubbery, flower garden and kitchen garden, Dickson arranged them according to their growth type: annuals, biennials and perennials. The plants were listed using their scientific names, and the catalogue contained a brief explanation of the Linnaean System, which by the 1790s was the dominant classification system used by botanists in Britain, although not necessarily by plant nurseries.

These literary and scientific trimmings gave the catalogue – which was otherwise simply a list of seeds and plants – the air of something that was about more than commerce. Dickson even admitted that his nursery was 'by no means... in possession of all the plants mentioned in this Catalogue', a strange assertion given that the book was ostensibly published to advertise his own company's wares. But as he explained, 'perhaps no Botanic Garden in Europe can boast of such a treasure. Our intention was to make a Catalogue of real use to the lovers of gardening, agriculture &c...; we are determined to increase our collection, and make it as complete as possible to supply the demands of the public.'[2]

---

[1] Thomson, *Seasons* (1735), 'Spring', ll. 247–250, quoted at the start of the Preface to Dicksons &co., *Catalogue* (1792), p. iii.

[2] Dicksons &co., *Catalogue* (1792), p. v.

Dickson's catalogue was distinguished from those of other British and French nursery companies by the inclusion of a well-known literary source. However, its use of Linnaean botanical nomenclature and its explanation of the Linnaean taxonomic system, although unconventional, were not completely novel.[3] Neither was its structure, nor its proposal of an inventory of 'all' interesting plants within Britain. Upperend plant traders like Dickson had experimented with the inclusion of such features over the preceding forty years or so, seeking to imply, or underline, an association between the commerce and science of plants. Archibald Dickson lived and worked in a world in which cultural motifs, scientific knowledge and trade were deeply intertwined.

This chapter explores that world. Botany occupied a prominent place in eighteenth-century British and French culture, and botanically interesting plants (particularly exotics) became objects of commerce. By marketing their plants in a particular way, the elite plant traders stimulated public interest in botany. We can trace this through the ways in which the traders presented themselves and their wares to consumers. Their advertisements, letters and catalogues, and even court cases concerning thefts from nurseries, reveal how they enlisted cultural and scholarly motifs to sell plants as scientific specimens. The ways in which they deployed such motifs affected, and in turn were designed to affect, the social place that the traders occupied and how the wider public understood botany as a science. The previous chapter examined the plant trade in terms of its social structure and the opportunities it presented. We now turn to the business culture behind eighteenth-century botany.

Plants and their Publics

Educated readers of Dickson's *Catalogue* would have recognised the excerpt quoted in the Preface as lines from James Thomson's celebrated poem *The Seasons*. Perhaps they would also have recalled the preceding passage, which dramatically portrays a botanist at work in the field. Thomson's botanist 'steals along the lonely dale / In silent search' of plants. He energetically 'bursts' through 'rank' forests, and finally he 'climbs the mountain rock, / Fir'd by the nodding verdure of its brow'. Yet despite his best efforts, the valiant botanist is outdone by the natural profusion that surrounds him, for 'With such a liberal hand has

---

[3] In contrast to commercial nursery catalogues, it was more common for horticultural books (which were didactic rather than simple lists) to include short literary quotations. Examples include: Society of Gardeners, *Catalogus Plantarum* (1730); Miller, *Gardener's Dictionary* (1752); Weston, *Universal Botanist* (1770–1777) and *Flora Anglicana* (1775). The mottos were all Latin quotations from Virgil, and were all on the title page.

NATURE flung / Their seeds abroad' that it is impossible to count everything.[4] Nature is fecund and 'innumerable', marvellous and beautiful to behold.

The plants that courageous botanists brought to Britain and France initially reached the wider public through visual or textual representations. The physical specimens themselves were very rare, and were mostly cultivated in private botanical collections. But the situation changed from the mid-eighteenth century onwards: thanks to the nursery trade, extraordinary flora found its way into more modest gardens, flowerpots and window boxes. Plant traders such as Dickson, Kennedy and Lee, Andrieux and Vilmorin, and their associates, were important because they translated the work of the 'intrepid botanist' for domestic audiences, and thus made the products of scholarly labour available for public consumption.

Who did the nurseries interact with, and how did they do so? This apparently simple question is complicated by the paucity of the remaining records. The surviving archival evidence about the nurseries mostly comprises letters exchanged between botanists – and not between the merchants themselves. Correspondents who mentioned the elite plant traders depicted them as go-betweens for two primary constituencies: on one hand, wealthy aristocrats and botanical institutions (a single group from the perspective of plant traders, because they often acted together as patrons); and on the other, a wider public of consumers. The traders re-articulated botanical knowledge, making it comprehensible to non-specialists among both these groups. Forming good relationships with each was central to the accumulation of information, to the exchange of both knowledge and specimens, and thus to the overall success of the plant nurseries. For the nurseries, however, the exchange of information was not straightforward. They were, after all, commercial businesses, and sharing knowledge (or specimens) too freely could potentially undermine their commercial interests. Getting it wrong – or giving away too much information to the wrong person – might jeopardise a nursery's scholarly status and its economic security.

While archival evidence is slight, the catalogues and other publications issued by the nurseries offer a way of ascertaining how the traders sought to present themselves in public. Ranging from two to two hundred pages in length, most of the fifteen English and Scottish catalogues discussed here were quite simply lists of plants, usually arranged according to the areas of the garden in which they could be cultivated.[5]

---

[4] Thomson, *Seasons* (1735), 'Spring', ll. 250–257.
[5] This sample includes all the traders' catalogues published between 1760 and 1800 that I have been able to locate. About half of these are digitised and available via Eighteenth

The French catalogues, including those issued by Andrieux and Vilmorin, were similarly laconic, revealing little about their authors or their envisaged readership.[6]

The catalogues gave little detail that related to trade, yet this was surely their *raison d'être*: it would be strange to go to the expense of compiling and printing these books unless they contributed to business. The nurseries' names and addresses were displayed prominently on the title pages, but in the British catalogues there were no further attempts to encourage potential consumers to contact them for their wares, and neither the French nor the British catalogues contained further discursive sections. In spite of their taciturnity, the catalogues are revealing of the status claimed by the merchant in question, and thus of the distinction between elite and common plant traders. By 1799, the upper-end nurseries had come to use their publications as a means of marking the difference between themselves and the majority of plant merchants. They used two key strategies to do so, which related to the way the plants were marketed and the language used to describe them. They also began to experiment with the catalogue genre, moving from simple trade lists to didactic texts more akin to books of horticultural instruction.

The vast majority of eighteenth-century plant catalogues did not print the prices of plants for sale. Leaving catalogues blank was partly a practical move: traders could annotate prices by hand without needing to have new lists printed every time something changed. Printing became more affordable over the eighteenth century, and in the 1770s four of the fifteen British catalogues began to include prices.[7] These four were all 'common' nurseries, however, and the upper-end traders persisted in publishing catalogues that left out the prices. Their unpriced catalogues served an additional function, beyond the practical consideration of saving on printing costs: they were intended to appeal to a group of consumers who might consider themselves to be 'genteel' and 'polite', and for whom engaging directly in commerce would appear rather unseemly. That the nurseries produced catalogues apparently downplaying their

Century Collections Online. The others are from the British Library (especially the Joseph Banks collection, where several catalogues had been bound together in one volume) and the Bodleian Library. Powell and Eddie's *North American Tree, Shrub and Plant Seeds* (c. 1764) is stored among the Malesherbes papers in AN, 399 AP 97.

[6] The French sample is much smaller than the British, because the French nursery trade was more localised and because the upper-end nurseries preferred to publish their catalogues within horticultural books, which will be discussed at more length later. The Andrieux and Vilmorin catalogues are: Andrieux, *Catalogue de toutes sortes de graines* (1760) and *Catalogue raisonné des plantes* (1771); Andrieux and Vilmorin, *Catalogue* (1778).

[7] Gordon *et al.*, *Catalogue* (1783); Goring and Wright, *Catalogue* (1798); Perfect and Perfect, *Catalogue* (1777); Telford and Telford, *Catalogue* (1775).

commercial nature may seem paradoxical, but it provides an insight into how the plant traders wished to present themselves, and who they imagined their public to be.

The trade in exotic plants fell squarely within the domain of luxury and semi-luxury consumption, and the practices used by both British and French traders to sell plants mirrored the methods used by other 'genteel' vendors. In this context, it was exceptional for prices to be displayed openly; they were negotiated at the very end of a customer's visit.[8] Vendors would then offer credit as a kind of gift that demonstrated that they trusted the customer to pay his or her debts eventually. The gift of credit consequently ensured consumer loyalty.[9]

Luxury shopkeepers fostered personal relationships with their customers, offering, for example, 'tea and conversation' to those who visited their premises.[10] In Paris, only the most prestigious merchants of any product could attract aristocratic clientele inside their shops. In the same way, Andrieux and Vilmorin encouraged customers to visit their nursery, where the proprietors would wait upon them so that they could 'see and choose, flower by flower' the plants in which they were interested.[11] Assuming that this strategy was successful, the nursery company would be ranked among the most elite Parisian luxury traders.[12]

The sale of plants was thus characterised by sociable encounters, which allowed prices to be established and which encouraged fidelity. Sociability also encouraged the exchange of knowledge between traders and customers. In these commercial contexts, behaviour was framed by the notion of politeness, which (as explained in the Introduction) defined 'correct' address and comportment, and which could be demonstrated through the ornamentation of one's self and one's shop (or garden, in the case of the nurseries).[13]

---

[8] Berry, 'Polite consumption', pp. 390–391.

[9] Nevertheless, reclaiming such debts was a complex business that required careful negotiation by the vendor. For Britain, start with: Hubbard, 'Art of advertising', vol. 2, pp. 274–275, 292–322; Finn, *Character of Credit*, pp. 8–11, 17, 90–91; Berry, 'Polite consumption', pp. 388–389; Smail, 'Credit, risk and honor', pp. 439–456. For France, start with: Coquery, *L'hôtel aristocratique*, pp. 21–22, 149–178; Crowston, *Credit, Fashion, Sex*, ch. 4.

[10] Berry, 'Polite consumption', p. 387; Finn, *Character of Credit*, pp. 89–90.

[11] SV, Vilmorin – Vieux Documents, Annonce (1782, 1786). Andrieux and Vilmorin encouraged people who could not visit the garden in person to correspond with them instead.

[12] Coquery, *L'hôtel aristocratique*, p. 79.

[13] Courpotin, 'De la boutique', pp. 320–328; Finn, *Character of Credit*, p. 21; Hubbard, 'Art of advertising', pp. 79–83, 89–100, 308–310, 314–322; Klein, 'Politeness for plebes', pp. 362–363; Walsh, 'Shopping et tourisme', p. 228.

Politeness alone does not explain why Archibald Dickson thought fit to include a literary quotation at the start of his catalogue. A second key characteristic of the upper-end nursery catalogues was that they used botanical nomenclature, a decision that contrasted greatly with the majority of plant sales catalogues, which used plants' vernacular names. Five of the British catalogues, all published between 1760 and 1792, deployed Linnaean nomenclature and classification, including Dickson.[14] Pierre d'Andrieux likewise used Linnaean nomenclature in his 1771 catalogue, and subsequent Andrieux and Vilmorin publications followed suit.

French eighteenth-century botany is often characterised as anti-Linnaean, because scholars working at the major Parisian scientific institutions (especially the Jardin du Roi) championed authorities such as Tournefort and Jussieu over Linnaeus. This was manifestly not the case for French botany more broadly, however, where there are plenty of examples of French advocates of Linnaeus.[15] Andrieux's decision to use Linnaean nomenclature and the Linnaean System would appeal, then, to customers across France, and, as he himself stated, would permit international trade as well.[16]

The elite nurseries' adoption of Linnaean nomenclature gestured towards their international outlook and eased communication with other botanists and gardeners – a fact that was celebrated widely among botanists and traders across Europe.[17] It also helped to underline a connection between their commercial activities and contemporary science. The British Linnaean catalogues that included a preface did so to assert their scientific understanding. John Webb, the first trader in Britain to publish a catalogue arranged according to a botanical taxonomy (in 1760), explained that he had 'placed [the plants in his catalogue] ... mostly under their present titles of the latest System of Botany formed by Dr. *Linnæus*'. He also offered a rationale for the selection of plants listed.[18] In 1778, William Malcolm claimed his sole purpose in writing his Preface was to 'recommend one regular and universal System of Botany ... the Sexual System of Linnæus'.[19] In 1783, partners Warren Luker and Samuel Smith emphasised that 'as it is equally

---

14 Dicksons &co., *Catalogue* (1792); Kennedy and Lee, *Catalogue* (1774); Luker and Smith, *Catalogue* (1783); Malcolm, *Catalogue* (1778); Webb, *Catalogue* (1760).

15 Duris, *Linné et la France*; Livesey, 'Botany'; Williams, *Botanophilia*.

16 Andrieux, *Catalogue raisonné des plantes* (1771), pp. i–ii.

17 Ibid., p. i; Lebreton, *Manuel de Botanique* (1787), pp. vii–ix, xiii.

18 Webb, *Catalogue* (1760), p. iii. The Linnaean System was not widely used in Britain before the 1760s. James Lee's *Introduction to Botany* (1760) made it publicly accessible for the first time, so Webb's was a state-of-the-art catalogue.

19 Malcolm, *Catalogue* (1778), p. iii.

necessary to know them [plants] Botanically, the Latin Botanic names are added for the convenience of the scientific', and they appended a Linnaean 'Generic Index' to their catalogue.[20] Publishing in 1792, Archibald Dickson was the boldest of all in pinning his colours to the scientific mast: 'The vast improvements Gardening...has received of late from the Science of Botany', he declared, have 'induced us to lay the following sheets before the public.'[21] In deploying Linnaean nomenclature, the traders demonstrated their understanding of cutting-edge botany – showing that the nurseries were *au courant* with the latest developments in science.

The third strategy used by the upper-end nurseries related less to catalogues specifically and more to attitudes towards publishing. The elite traders and gardeners diversified the kind of publication that they issued, experimenting with the production of practical manuals that dealt with both horticulture and botany. They used their publications to emphasise their intellectual prowess: James Lee's *Introduction to Botany* featured classical references, for example. He described how Linnaeus had 'led your Readers, by an *Ariadne's* Clue', and he construed contemporary discussions about botany in terms of the debate between Ancients and Moderns. 'Accordingly', Lee declared, 'we find, in the Account given by *Herodotus*...'[22] Such a reference was surely designed to draw the author closer to potential customers from the genteel classes. Archibald Dickson's use of Thomson's poem epitomised the same aim. These upper-end nurserymen demonstrated that they were not only educated in botany but had also received a broad classical and literary education. Their cultural capital facilitated communication with the genteel social groups whose custom – and patronage – they desired.[23]

Introducing the Amateur

The diversification in genres was much more evident in France than in Britain, and it took a particularly Gallic form, as the plant traders presented themselves not just as merchants, but as members of a group of enlightened 'amateurs'. The culture of amateur botany is the subject of

---

[20] Luker and Smith, *Catalogue* (1783), p. v.   [21] Dicksons &co., *Catalogue* (1792), p. iii.

[22] Lee, *Introduction to Botany* (1765), pp. iii, vi. James Lee could probably also read Latin. The *Introduction to Botany* (1st edn, 1760) was a loose translation of Linnaeus' *Philosophia Botanica*. He wrote to Linnaeus in English, but received replies in Latin. See: LS, L5238, James Lee to Carl Linnaeus, 4 October 1776, and LS, L5239, Carl Linnaeus to James Lee, November 1776.

[23] Cohen, *Fashioning Masculinity*, pp. 47–50.

the next chapter, but we should note here that in France an amateur was not simply an *honnête homme* (the French equivalent of a gentleman);[24] he or she was someone who purportedly possessed a 'natural' ability to cast judgements, and the authority to act as an arbiter of taste. Taken together, the various elements of amateurship offered a cultural model of refinement, education, fashion and tastefulness.[25]

It was common for French upper-end nurseries, such as that run by Andrieux and Vilmorin, to have their catalogues printed within books on horticulture in addition to the catalogues they issued independently. The horticultural books were usually cheap octavo publications of between one and four hundred pages. They contrasted with the trade catalogues discussed earlier because they blurred the boundaries between horticulture, botany and trade, as their titles – *Bon Jardinier*, *Jardinier Portatif*, *Traité des Jardins* – might suggest.[26] In addition to including practical instructions for the cultivation of plants, the texts recommended learning botany as a counterpart to gardening and included lists of the plants that readers could purchase from specific nurseries. The traders also contributed essays about horticulture to these books. The publications thus endorsed the hybrid knowledge that the upper-end plant traders had mastered.

The horticultural books were edited (and largely written) by individuals such as Thomas-François Le Grace (1713–1798), a former royal censor and published author of treatises on history, agriculture and horticulture, and René Le Berryais (1722–1807), a noted agricultural reformer and collaborator with Duhamel Du Monceau.[27] The editors explicitly identified themselves as amateurs, either on the title page or in the Preface of the book.[28] They also emphasised that they were active participants in the public sphere, asserting that their actions were

---

[24] *Honnêteté* was not a direct parallel of British gentlemanliness. For more on *honnêteté* and the *honnête homme*, see: Cohen, *Fashioning Masculinity*, pp. 14–16; Crow, *Painters and Public Life*, p. 67.

[25] Guichard, *Les amateurs*, p. 17. See also: Coquery, *La boutique et la ville*, Présentation, p. 10.

[26] The *Bon Jardinier* (discussed in Chapter 1, p. 58) doubled in size between 1768 (208 pages) and 1785 (456 pages). The editions discussed here are those published in 1768, 1773 and 1785. All editions cost 36 sols relié.

[27] Grand-Carteret, *Les Almanachs Français*, p. 66; Quérard, *La France Littéraire*, p. 12.

[28] The 1768 and 1773 editions of the *Bon Jardinier* were both apparently 'entièrement refondu par un Amateur' and the instructions in the *Jardinier Portatif* (1772) were 'exactement expliquée par un AMATEUR' (emphasis in the original). The author of the 1785 edition described himself as a 'Censeur Royal, Amateur & Cultivateur'. The authors of the *Traité de la Culture de Différentes Fleurs* (1765) and of the *Traité des Jardins* (1789) didn't describe themselves as amateurs but explained that this was their envisaged readership.

supported by a wider community of readers and correspondents.[29] The *Avertissement* of the 1773 edition of the *Bon Jardinier* thus announced that 'the welcome that the public has given to the almanac *le Bon Jardinier*, has engaged the author to profit from the observations that Connoisseurs...have made'.[30] This sense of community was reinforced further by frequent references to things known by 'tout le monde', or 'everyone'. For example, in its discussion of the 'aloès uvaria' (*Aloe uvaria*, L.), the *Bon Jardinier* explained that 'Everyone knows that one should give them but very little water during the summer', and concluded that a particularly uncommon and leafy species of this plant 'deserves to enter into the garden of *un Curieux*'.[31] The rhetoric of amateurship suggested that the author's advice and actions had been approved by the wider public.[32]

The 'amateur' editors also drew hierarchical distinctions between themselves and botanists, presenting the latter as authorities responsible for determining botanical taxonomy and nomenclature. The implication of this restrictive definition was that a member of the wider public who studied botany could not claim authority as a 'botanist'. In 1785, the editor of the *Bon Jardinier* thus cheerfully stated that he would name a plant newly arrived from China 'the *Aerial plant*, just until *Messieurs les Botanistes* have given it a name.'[33] The authors and editors of these books positioned themselves as people who, like their readers, did not share the taxonomic concerns that preoccupied recognised authorities on botany.[34]

Plant traders presented themselves as people who could move between both arenas, and who could thus translate science for a wider audience. Like the 'amateur' editors of the *Bon Jardinier*, British and French traders carefully distinguished themselves as different from botanical authorities, whose primary concern was with establishing botanical

---

[29] In the same way, some medical physicians constructed their authority as practitioners by building a discursive relationship (in print) with a wider 'public'. Broman, 'Habermasian public sphere', pp. 133–139.

[30] Anon., *Bon Jardinier* (1773), p. xvii. On the relationship between the 'connoisseur' and 'amateur', see Chapter 3, esp. p. 81.

[31] Anon., *Bon Jardinier* (1773), p. 93.

[32] Broman, 'Habermasian public sphere', pp. 133–139, 144; Shapin, *Social History of Truth*, pp. 36–41. This rhetoric of amateurship was also evident in the correspondence between gardeners. See: LS, L5601, Thomas Knowlton to John Ellis, October 1770. See also: Hilaire-Pérez, 'Diderot's views', pp. 146–147 and *L'Invention Technique*, pp. 301–303.

[33] de Grace, *Bon Jardinier* (1785), p. 419.

[34] Chapter 3 discusses at length the place of taxonomy within the construction and practice of amateur botany.

systems. The position taken by plant traders was made very clear at a court trial in London in 1795, when three nurserymen were cross-questioned about the extent of their botanical knowledge. Nurseryman James Colvill was described as a 'student of botany', and asserted that studying the science 'has always been my delight'. But he was not described as a 'botanist'. John Frazier was more circumspect about his knowledge. Despite the fact that, as he explained, 'I have travelled upwards of fifty thousand miles in search of plants ... [and] have discovered above three hundred new species and fifty new genera,' when asked whether he considered himself 'to have a competent skill in Botany', he replied, 'By no means; I do not'. Even though he clearly understood the science of plants, he was unwilling to claim the title of 'botanist'. Nurseryman Charles Scoby was differentiated from the first two witnesses because he did 'not deal in rare plants'; his stock was 'the common sort'. When asked whether he had 'made botany your study', he replied, 'A little; I am not what they call a proficient, nor do not [sic] pretend to it'.[35] Thus, while British upper-end plant traders were keen to associate themselves with eighteenth-century scholarly culture, and to underline that they knew botany, they did not claim the designation 'botanist'. Their rejection suggests that, as in France, 'botanists' were perceived as a discrete group of experts within a wider culture of learning.

Nurserymen Pierre d'Andrieux and Philippe-Victoire Lévêque de Vilmorin were unusual compared to most other French plant traders because each was officially titled 'botaniste du roi'. This royally granted designation suggested superiority – and authority – over the wider botanical and horticultural community. They negotiated this by carefully associating themselves with the positive image of the amateur as elaborated in books such as the *Bon Jardinier*. Andrieux and Vilmorin presented themselves as socially respectable and erudite, and adopted the model of the 'enlightened' amateur who intended his or her scholarly activities to be of benefit to the rest of society. Cultivating these associations added a positive sheen to the marketing of their goods.

The creation of a collection was particularly important within the culture of amateurship, demonstrating the proprietor's ability to select and arrange the objects of their study in a refined manner.[36] Visitors acted as a 'tribunal of taste', as explained by amateur of art Pierre-Jean Mariette (1694–1774) in 1750:

---

[35] Old Bailey Proceedings Online (OBP) (www.oldbaileyonline.org, version 7.2, 11 August 2016), 16 September 1795, trial of Charles Fairfield (t17950916-73).

[36] Guichard, *Les amateurs*, pp. 15–17, 158–159; Pomian, *Collectionneurs*, p. 158.

It is in the choice of [art] works that an Amateur makes his discernment known, and [this is] how he shows whether he has taste, or whether he is deprived of it. His Cabinet is, as it were, a tribunal where we judge him without mercy.[37]

Like the cabinet described by Mariette, collections of live plants in private gardens were open to a select group of amateur visitors. Andrieux and Vilmorin made much of this, ensuring that the public understood that their collections were part of this circuit. They thus announced that their gardens were open for the enlightenment of, and judgement by, any interested person, and they corresponded with the wider learned community. The *Bon Jardinier* and the *Traité des Jardins* printed the names and addresses of Andrieux and Vilmorin, as well as several amateurs located in and around Paris. Readers were told that if they visited these gardens, they could obtain more information and advice about the plants that they cultivated or wished to cultivate. The succession of commendations effectively created a ready-made horticultural itinerary around Paris.

Suitable gardens were also regularly recommended to amateurs as places of botanical interest in guidebooks such as Thiéry's *Guide des amateurs*. Several of these are shown on Map 1.2. The garden of M. de Saint-Germain in the Faubourg Saint-Antoine (M on Map 1.2) was a particular favourite, as was that of Jacques-Martin Cels at the Barrière St Jacques (O on Map 1.2).[38] Cels' collection was highly regarded; the range of rare plants he cultivated reflected his acknowledged botanical expertise and his extensive scholarly connections. He was also one of only sixteen people who possessed a key to the Botany School at the Jardin du Roi in the 1770s.[39] 'Amateur' readers of the guidebooks and journals were explicitly encouraged to visit these private gardens. M. de Saint Germain, for example, emphasised that 'it would give him pleasure to show [his garden] . . . to Amateurs'.[40]

Self-described amateur gardeners used journals to publicly announce news about the rare specimens they possessed. On 11 July 1783, M. de Saint-Germain announced in the *Journal de Paris* that 'the superb flower of the *Cactus grandiflorus*' had opened: 'the most beautiful and the

---

[37] Mariette, *Traité des pierres gravées* (1750), in Pomian, *Collectionneurs*, p. 158 (my translation).

[38] De Grace, *Bon Jardinier*, p. 195. Cels worked as a *receveur* at the Barrière St Jacques and maintained a garden there from at least 1778 until 1791. He had moved to Montrouge by 1792. See: MNHN, Ms. 314, État de la Correspondance d'André Thouin; BL, Add. Ms. 8098, Cels (Paris) to Joseph Banks (London), 29 March 1792; BL, Add. Ms. 8097, ff. 393–394, Cels (Paris) to Joseph Banks (London), 7 March 1791. The contents of Cels' Montrouge garden are listed in a post-mortem inventory made in 1808. AN, ET/XLIV/0765, Inventaire après décès de Jacques Martin Cels, 27 June 1808.

[39] Spary, *Utopia's Garden*, pp. 56–57.   [40] de Grace, *Bon Jardinier* (1785), p. 195.

biggest of all known flowers'.[41] Publications such as the *Journal de Paris* also acted as forums for public conversations about the plants cultivated in these collections. In 1785, for example, readers discussed the *Sophora japonica*. Contributing to the conversation, the duc de Noailles' librarian wrote that '[t]he *sophora* of Japan flowered for the first time in 1779, in the gardens of the Maréchal de Noailles. [It flowered n]ext in 1781 and two years later, in 1784.'[42]

The gardens owned by both amateur collectors and commercial nurseries were given public prominence as repositories of rare plants and were presented as essential locations where other scholars could view and study the rare and unusual.[43] The French plant traders did not necessarily aspire to be amateurs, but they used their publications to demonstrate that they understood, and could therefore support, the refined amateur culture of their customers. In Britain, the upper-end plant traders did not adopt the model of amateur scholarship as explicitly as their French counterparts, but the image that they sought to project in public was nevertheless similar in terms of its scholarly pretensions. On both sides of the Channel, then, the traders neither claimed to be scientific authorities nor asserted that they were gentlemen. Nevertheless, they used their own nursery collections to exhibit their expertise and their tasteful discernment. In France, this was equated with amateur culture, and in Britain, with notions of polite gentility.

In spite of the different appellations, the upper-end French and British plant merchants thus acted in very similar ways. Both built up reputations for their philanthropy and for their generosity in sharing specimens and information. Andrieux and Vilmorin were celebrated for their willingness to give away specimens for free,[44] and James Lee 'might have died rich...but he was notoriously generous, and cared not what expenses he was at for the attainment of rare plants'.[45] Such displays of liberality were important because they strengthened the association between the nurseries and the public good, elevating them above trade and profit and bringing them closer to public-spirited scholarly culture.

The parallels between the British and French forms of self-presentation are testament more to the transnational continuities of

[41] *Journal de Paris*, no. 192, 11 July 1783. I am grateful to Professor Bruno Belhoste for this reference.

[42] Noailles' garden is K on Map 1.2. *Journal de Paris*, no. 257, 14 September 1785. I am grateful to Professor Bruno Belhoste for this reference.

[43] On the seventeenth- and early eighteenth-century precedents to this, start with: Goldgar, *Tulipmania*, pp. 78–81 and Coulton, 'Curiosity, commerce and conversation'.

[44] Heuzé, *Les Vilmorin*, pp. 13–14; Silvestre, 'Notice Biographique', pp. 208–210.

[45] Thornton, 'Sketch of the life of James Lee' (1810), pp. xv–xvi; Willson, *James Lee*, p. 31.

Enlightenment culture than to equivalences between the different commercial contexts. This is especially evident because the plant traders
espoused many of the principles that underlay Enlightenment scholarship. They reinforced their affiliations with scholarly culture by sharing
practical knowledge as well as specimens, for example. The response
from the wider scholarly community was generally very positive. Writers
on agronomy, for example, recognised that the plant traders' accrued
skill and knowledge could be applied not only to the perfection of
gardening, but also to agricultural improvement. British agricultural
reformer Richard Weston (1733–1806) studied 'the methods practiced
by the most skilful nurserymen, and London kitchen-gardeners' and
concluded that 'husbandry will never arrive at half the perfection that it
is capable of, till the garden-culture is more imitated in the field'.[46] The
munificent gestures of these enlightened merchants provided prominent
models of praiseworthy amateurship, showing that their personal expenditure had direct public utility.

The form of the nursery catalogues and the emphasis placed on the
scientific and practical knowledge of the plant traders also suggest that
the culture surrounding how plants were purchased had changed in
important ways. In wealthier households in early eighteenth-century
Britain and France, the responsibility for ordering plants had fallen to
the head gardener or steward.[47] For the middling sort, decisions about
planting and orders for plants were made by either the master or the
mistress of the house, perhaps in consultation with a hired gardener or
with the local plant nursery.[48] We might conclude that these groups were
the primary target audiences of the nursery catalogues, but the more
refined plant traders instead presented themselves as 'genteel' traders,
and assumed that their readers would share their polite perspective on
the commerce of plants. As gardening became more widely practised,
the responsibility for ordering plants, and therefore the opportunities to
communicate with educated plant traders, broadened out from the gardeners and stewards of great households to include their masters and
mistresses. Men and women of more modest means also increasingly

---

[46] Weston, *Tracts* (1773), pp. iii, xi.

[47] In 1802, the gardener Thomas Dixon advised a young friend how to order seeds from a
nursery for his employers: 'write them down in your letter [to the nursery] one sort after
another as you see them in the Seed bill without mentioning the prices – and you can
write them first over and let your M^{rs} see the order – then if there be anything mentioned
they do not want...you can write it over again to their mind'. CRO, D/Sen/5/5/1/9,
Thomas Dixon (Netherby, Cumberland) to John Johnston (Netherhall, Cumberland),
16 December 1802.

[48] Longstaffe-Gowan, *London Town Garden*, p. 151.

took up gardening as an agreeable pastime.[49] So the traders did not direct their catalogues only to stewards. They knew, or imagined, that as interest in botany and horticulture was spreading through society, an increasing number of property-owners were taking a personal interest in selecting specimens for purchase. Thus, it made sense to demonstrate a strong affiliation with polite or *honnête* culture.

British and French plant merchants consequently presented themselves as relatively socially elevated. While the British wished to be seen as 'genteel' traders, their French counterparts took this one level further, suggesting that they, too, were amateurs of nature.

Plant Breeding and Market Competition

At the heart of this polite self-presentation, however, we find a contradiction: the nurseries' claims to scholarly status were juxtaposed against their activities as traders. Operating within an increasingly competitive marketplace, the elite nurseries needed to assert their distinctiveness in relation to their rivals while simultaneously upholding their claims to scholarly munificence. The former appears to negate the latter. The elite nurseries were dedicated to the importation and naturalisation of exotics and the breeding of varieties for horticultural and agricultural use. But did they – *could* they – assert their 'ownership' over these specimens? The relationship between plant breeding and intellectual property has been mainly explored for the nineteenth and twentieth centuries; very little research has been done for periods before the advent of genetics, and when the cultural meaning of 'authorship' was very different to the present day.[50] To what extent did eighteenth-century plant breeders consider their productions to be their 'property'? And what impact did this have upon their position in the marketplace, and particularly on their claim to membership of Enlightenment scholarly culture? We can answer these questions by returning to the court case briefly mentioned earlier.

On 16 September 1795, the English gentleman Charles Fairfield was in court. Fairfield, a painter by profession and amateur plant collector, had been accused by nurseryman Daniel Grimwood of having stolen several prized exotic plants from the hothouse on his premises. Fairfield pleaded his innocence, but despite calling forth 'students of botany' as

---

[49] Harvey, *Early Gardening Catalogues*, p. 46.

[50] Bonneuil, 'Mendelism'; Burgos and Kevles, 'Plants as intellectual property'; Gaudillière *et al.*, *Living Properties*; Kevles, 'Patents, protections, and privileges'; Nelson, 'Is there an international solution'. An earlier version of my discussion of eighteenth-century attitudes towards plant breeding is included in Gaudillière *et al.*, *Living Properties*, as Easterby-Smith, 'Propagating commerce'.

expert witnesses, the judges struggled to ascertain any solid proof to support the claims made by either prosecution or defence. The problem they faced was how to differentiate positively one plant of the same species from another, and thus how to prove who was the rightful owner of the allegedly stolen goods. Indeed, as one witness put it, 'There is no man on earth can say that [they can tell one specific plant from another], for there is a likeness of every plant on the face of the earth, of the same species and genera.' Charles Fairfield was consequently acquitted.[51] But one way in which plant traders could, and did, introduce potentially identifiable difference was through breeding their own varieties of plants. Had Grimwood done this with the plants in question, it might have been easier for him to argue his case – or then again, perhaps not. For how did breeders prove that varieties were 'theirs', and to what extent did they establish legal title to ownership in the later eighteenth century?

The eighteenth century saw extensive interest emerge in developing new varieties of plants; an interest that was shared by plant traders, amateurs and patrons of botany and that accelerated over the period. As early as 1716, nurseryman Thomas Fairchild (1667–1729) had created the first recorded man-made hybrid by crossing a carnation with a sweet william, to produce the 'Fairchild Mule'. This was a significant step in the history of plant breeding, opening up the possibility of endless experimentation.[52] Knowledge about vegetable physiology then developed rapidly, thanks to the efforts of botanists and gardeners such as Stephen Hales, whose groundbreaking experiments on plant physiology were published in his book *Vegetable Staticks* in 1727. Hales' work was translated into French by Georges-Louis Leclerc (later the Comte de Buffon) in 1735, and inspired further research among gardeners and botanists.[53]

A 'variety' was usually defined as a plant or animal that differed in a minor way from the species to which it belonged.[54] Variation could be produced in two ways, either naturally or through human intervention.[55] Varieties intentionally produced through cultivation, which are now known as 'cultivars', were rarely considered worthy of

---

[51] OBP, 16 September 1795, trial of Charles Fairfield (t17950916–73); Cust, 'Fairfield, Charles'.

[52] Wulf, *Brother Gardeners*, pp. 6–16.

[53] Bungener, 'La Botanique', pp. 295–296; Henrey, *Botanical and Horticultural Literature*, vol. 2, p. 644; Roger, *Buffon*, pp. 25–27.

[54] 'Variety' in OED Online, www.oed.com/view/Entry/221557 [accessed 11 August 2016]; UPOV, 'The notion of breeder'.

[55] Linnaeus explained that 'varieties are plants changed by some accidental cause'. Linnaeus, *Philosophia Botanica*, Aphorism 158, in Stafleu, *Linnaeus*, p. 90. See also the discussion of varieties of the 'Angoulême Pear' in Blaikie, 'Foreign notices' (1828).

botanical attention.[56] Instead, they were presented as 'sports' of nature, produced to satisfy human desires.[57] The production of varieties intersected with the culture of horticultural collecting rather than with taxonomic botany, and was prominent within the ornamental plant trade. As early as 1629, John Parkinson gave detailed descriptions of four varieties of *Aconite*, and explained that, in addition, 'Many more sorts of varieties of these kinds there are, but these onely... are noursed up in Florists Gardens for pleasure; the other are kept by such as are Catholicke observers of all natures store.'[58] A 'hybrid', in contrast, was defined in 1775 as 'produced from plants of different kinds', or 'begotten between animals of different species'.[59] Although these 'vegetable mules' were undoubtedly 'an irrefragable argument in favour of the sexual system of botany', botanists saw them as even less useful to science than cultivars.[60] In the fifth edition of his *Introduction to Botany* (1794), James Lee declared that a hybrid was 'a Bastard or Monstrous production'.[61]

Skilled gardeners and plant traders were, nevertheless, actively involved in breeding a great range of new plants. The flora available for cultivation diversified rapidly, and garden designs were soon altered to include more colour and variety.[62] Evidence about how either Kennedy and Lee or Andrieux and Vilmorin actually conducted their experiments is frustratingly limited, yet their surviving correspondence shows that they were successfully developing hardier varieties of exotics that could survive in European climates.[63] Surprisingly, however, neither nursery gave its breeding activities much publicity. The catalogues they produced between the 1770s and 1790s identified plants by genera and species, but rarely specified precisely which variety was for sale. Indeed, the evidence furnished by both nurseries suggests that, at this time, breeders

---

[56] The word 'cultivar' was invented in 1923 to signify a 'cultivated variety'. 'Cultivar' in OED Online, www.oed.com/view/Entry/45722 [accessed 11 August 2016].

[57] Mortimer, *Whole Art of Husbandry* (1721), vol. 2, p. 217.

[58] '*Aconitum*. Wolfebane', in Parkinson, *Paradisi in Sole Paradises Terrestris* (1629), ch. xxvi, p. 215.

[59] 'Hybrid', in Ash, *New and Complete Dictionary* (1775).

[60] Darwin, *Botanic Garden, Part II* (1789), p. 149.

[61] 'Hybrida', Lee, *Introduction to Botany* (1794), p. 418; Stafleu, *Linnaeus*, pp. 90–91.

[62] Bourret *et al.*, *Parcours*, p. 192ff.; Laird, *Flowering, passim*, esp. pp. 13–16 and *Formal Garden*, pp. 91–95; Wulf, *Brother Gardeners*, pp. 7–8.

[63] For examples, see: BL, Add. Ms. 33540, f. 260, James Lee (London) to Colonel Bentham (Constantinople?), 14 June 1786; BL, Add. Ms. 29533, f. 59, James Lee jr. (London) to [unknown correspondent], 14 December 1796; AN, 399/AP/99, Vilmorin and Andrieux (Paris) to Malesherbes, 3 November 1782 and 17 November 1782; BNF, NAF 2757, ff. 26–27, Lullin (Geneva) to Vilmorin-Andrieux (Paris), 23 September 1806; BNF, NAF 2758, f. 30, Louis Ordinaire (Belfort) to Vilmorin (Paris), 17 September 1813; BNF, NAF 2758, f. 129, Dumont de Courset (Pas de Calais) to Vilmorin (Paris), 14 August 1810.

were most concerned to advertise the new plants they had obtained from overseas, rather than to promote the new varieties that they had bred themselves.

Well-connected nurseries such as Kennedy and Lee and Andrieux and Vilmorin were at the top of the chain for receipt of new plants from plant hunters, but even so James Lee hired a boy to collect the empty seed packets that Philip Miller threw into the Thames from Chelsea Physic Garden. Obtaining these allowed Lee to monitor exactly which new plants Miller had received; he could also always hope to find one or two seeds left inside.[64] Andrieux and Vilmorin participated in extensive experiments to naturalise exotic plants. They shared seeds and discussed the results of their experiments with botanists and other scholarly cultivators in and around Paris, including Malesherbes, the Abbé Nolin (director of the royal plant nurseries) and André Thouin.[65] But although they were heavily involved in breeding hardier varieties, they did not assert that the varieties they produced were their 'own'. Access to new specimens as they arrived from overseas was essential to nurseries' commercial success; developing new varieties was secondary to this.

It is possible to determine attitudes towards property and ownership from the ways in which plant breeders named and described their productions.[66] Kennedy and Lee's seventy-six page *Catalogue* of 1774 listed exotic or rare ornamental plants, arranged according to hardiness.[67] The authors described the plants for sale using both Linnaean binomial nomenclature and the English vernacular or 'trivial' name. Like the other catalogues discussed earlier, this catalogue gave no further descriptive information about the plants. The lack of information seems especially surprising because many of these exotic species were probably unknown to readers.

In contrast to the English nursery's publications, Andrieux and Vilmorin did place emphasis upon plants that were new or rare, and they also specified some of the varieties on offer, although they did not identify which were the cultivars that they had bred.[68] Rather than simply

---

[64] Thornton, 'Sketch of the life of James Lee' (1810), pp. xv–xvi.

[65] For examples, see: AN, 399/AP/99, Vilmorin-Andrieux (Paris) to Malesherbes, 3 November 1782; AN, 399/AP/99, Vilmorin-Andrieux (Paris) to Malesherbes, 17 November 1782; AN, 399/AP/101, Vilmorin (Paris) to Malesherbes, 30 April 1792; AN, 399/AP/101, Draft letter from Malesherbes to Vilmorin, n.d. [1790s?]; BL, Add. Ms. 8096 f. 121, L'Héritier de Brutelle (Paris) to Sir Joseph Banks, 2 April 1785. See also the correspondence in AN, AJ/15/511, 'Envois de Graines, plantes, minéraux etc de pays étrangers au jardin du roi'.

[66] On the social history of botanical plant names, start with Schiebinger, *Plants and Empire*, ch. 5.

[67] Kennedy and Lee, *Catalogue* (1774).   [68] Andrieux and Vilmorin, *Catalogue* (1778).

listing the names of plants, their 1778 catalogue also included a sentence or short paragraph explaining the differing features and aesthetic impacts of the plants advertised. The entry for the 'Aconit Napel, *Aconitum Napellus*' (Aconite) explained that it had blue flowers and that it flowered in the summer. It listed one additional variety, the 'Tue-loup, *Aconitum Lycocotonum*' (Wolf's Bane), and explained that these were 'large vivacious plants, which grace *jardins anglais* very well.'[69] However, the inclusion of descriptions did not extend to adding Andrieux and Vilmorin's signature to either variety. As stated in the *Avis* at the start of the book, the intention behind providing the short descriptions was to elevate the catalogue to being 'more than a list'. The authors explained that a list 'would appear to be made less from the desire to be useful, than by the interest of making known the objects of our Commerce'.[70] The nursery's concern to ensure that it did not appear purely commercial is central to explaining why it and other nurseries refrained from adding their names to cultivars.

This lack of concern about specifying the differences between cultivated varieties was also reflected in correspondence between leading botanists throughout the 1770s and 1780s. Of Joseph Banks' correspondents who requested seeds, the majority asked only for the generic name, and rarely specified a particular type. In 1777, for example, the Danish botanist Peter Wilhelm Edinger asked Banks to obtain agricultural seeds for him from the nurseryman James Gordon. Edinger asked for: 'clean hayseed . . . Sainfoin . . . rye-grass [and] . . . yellow Clover'.[71] Yet, botanists and plant traders were well aware of the differences between certain varieties, and in other cases could be very specific. This was especially so in the context of the cultivation of older or more common flowers, fruits and vegetables, where the ability to produce variations had long been celebrated: 'Is not a unique Hyacinth', exclaimed George Voorhelm in his 1752 *Traité sur la Jacinte*, 'which twenty or thirty people have been trying in vain to cultivate, a wonderful thing?'[72] Andrieux and Vilmorin emphasised that they could provide 'toutes sortes', or 'all varieties', of

---

[69] Andrieux and Vilmorin, *Catalogue* (1778), p. 36. The '*jardin anglais*', or '*jardin anglo-chinois*', was a fashionable garden design in late eighteenth-century France. In contrast to the straight lines that had characterised French garden design since the late sixteenth century, the new 'English' style favoured curves, which were supposedly more 'natural'. The *jardin anglais* created new spaces in which to display non-native plants and trees, as well as exotic architectural features such as Chinese pagodas. For more on the *jardin anglais*, start with: Barrier *et al.*, *Aux jardins de Cathay*; Laird, *Formal Garden*.

[70] Andrieux and Vilmorin, *Catalogue* (1778), *Avis*.

[71] BL, Add. Ms. 8094, f. 154, Letter from Peter Wilhelm Edinger (Copenhagen) to Joseph Banks, 28 November 1777.

[72] Quoted in Goldgar, 'Nature as art', p. 338.

each of the flowers they advertised, and Joseph Banks also discussed at length different varieties of fruits and vegetables with fellow horticulturalist Thomas Andrew Knight (1759–1838).[73] Botanists and upperend plant traders observed and understood the differences between plant varieties, yet discussions of these were often downplayed.

The lower-ranking plant traders, who generally grew better-known plants such as native fruits and vegetables and 'florists' flowers' were not engaged in the same scientific culture as the more elite nursery companies. They consequently articulated different attitudes towards plant breeding and divergent conceptions of property, which included publicly promoting the varieties they had bred.[74] A catalogue of gooseberries sold by non-elite Lancashire growers in 1780 provides a pointed example. Although published as an appendix to a Linnaean English flora, the fruit were only listed according to their trivial, rather than botanical, names. In direct contrast to the catalogues discussed earlier, three quarters of the 319 gooseberry varieties were named after the cultivator. For example, the catalogue contained not only gooseberries named 'Red Mogul', 'White Mogul', 'Robin Hood' and 'Little John', but also 'Mather's White Mogul' and 'Thorp's White Mogul'. 'Withington' grew 'Little John', and 'Shuffleton' and 'Worthington' both grew their own versions of the 'Emperor of Morocco'. The gooseberry known simply as 'Seedling' was even more widely bred, with fourteen different breeders adding their own surnames to this variety. As the catalogue had neither illustrations nor written descriptions, the cultivators gave their plants memorable names, in particular choosing ones that evoked sentiments of triumph, heroism or exoticism.[75] In direct contrast to the upper-end plant traders, the cultivators of native or common plants sought to attract custom by associating their own names with specific plant varieties.

Plant Names and Botanical Networks

Historians of plant breeding have shown that market competition in the nineteenth century encouraged breeders to place more emphasis on their

[73] NHM, DTC, vol. 19, f. 300, Joseph Banks (Spring Grove) to Thomas Andrew Knight, 21 July 1816. Bigarreau cherries are red, pink or white, with a firm skin and sweet taste.
[74] Duthrie, 'English florists' societies', pp. 17–35; Nelson, 'Dublin Florists' Club', pp. 142–148.
[75] Maddox, 'Catalogue' (1780), pp. 115–118. According to William Withering, there were actually only two species of gooseberry grown in Britain at this time. See: Withering, *Arrangement of British Plants* (1796), vol. 2, pp. 266–267. See also: Abercrombie, *Garden Vade-Mecum* (1790), pp. 217–224. On the connections between botanists and gooseberry-growers in the nineteenth century, see: Secord, 'Science in the pub', pp. 276, 285.

'ownership' of the varieties that they produced. In doing so, breeders had to use botanical descriptions in order to identify plants as 'theirs'.[76] Yet, in the eighteenth century, the breeders whose education and social connections placed them closest to scientific knowledge seemed to care the least about asserting their ownership over the specimens that they produced. Furthermore, the upper-end nurseries rarely, if ever, actually specified the types of cultivars that they had bred, even though they possessed the knowledge necessary to identify and describe these differences. This apparent contradiction was due to the cultural conventions surrounding claims of 'authorship' (or, in modern terms, intellectual property) over a natural production.

The elite nurseries could obtain new exotic specimens because they maintained close connections with the botanical network. Plant hunters, many of whom had been trained in nurseries but were employed by esteemed botanists, brought exotic plants back to Europe. Nurseries' access to these new specimens depended crucially on their ability to integrate themselves within the scholarly community. Both Kennedy and Lee and Andrieux and Vilmorin did this very successfully: many of the plant hunters who sent either gifts or specimens to Joseph Banks asked him to forward part of the parcel to James Lee,[77] while Philippe-Victoire Lévêque de Vilmorin similarly received new specimens via the Jardin du Roi and from other 'botanical travellers'.[78] The nurseries were indispensable to botany, for they provided an additional location for the cultivation of new exotics, in which botanical expertise was combined with a high level of practical experience.

Plant breeders depended on their membership of the scholarly community in order to obtain new specimens that they could eventually sell. As outlined in the Introduction, admission to the Enlightenment scientific community of the 1770s and 1780s was, in principle, possible for

---

[76] Kevles, 'Patents, protections, and privileges', pp. 323–331.

[77] For example, in 1767, the Rev. John Walker told Banks that, 'I have put up for our Friend Mr. Lee at Hammersmith a small Parcel of Plants'; in 1783, the botanist Thomas Lauth sent two catalogues of the Strasbourg botanical garden to Banks, and asked him to give one to Aiton and the other to Lee; in 1784, Francis Masson sent parcels to Joseph Banks 'to be divided between Mr Aiton and Mr Lee'. See: Add. Ms. 35057, f. 24v, Rev. John Walker (Moffat) to John [i.e. Joseph] Banks (London), 28 March 1767; Add. Ms. 8095, f. 196r, Thomas Lauth (Strasbourg) to Sir Joseph Banks, 1 February 1783; PSJB, 5.13.10, Francis Masson (Puerta de la Orotava, Tenerife) to Joseph Banks, 19 March 1778; PSJB, 5.13.21, Francis Masson (Madeira) to Joseph Banks, 25 March 1784.

[78] For examples of such arrangements, see: MNHN, Ms. 1945, La Billarderie, Draft memo re: the voyage of Aristide Dupetit-Thouars, 31 August 1792; BNF, NAF 2758, f. 26, Abbé Nolin to M. le Consul de S . . . , September 1781. For more on the relations between botanists and botanical collectors, start with Spary, *Utopia's Garden*, ch. 2.

any scholarly practitioner of science. But this is not to say that obtaining membership was straightforwardly dependent on scientific knowledge and skill. Practitioners in both countries had to present themselves in ways that made them worthy of respect from other scholars. In particular, it was important to demonstrate their commitment to the 'public good'.[79] Traders' private interests could be construed as counter to this, because they aimed to make a personal financial profit from their scholarly activities.[80] In France, misgivings towards traders' self-interest were further reinforced because, although mercantile activity was no longer forbidden to the nobility, it was still viewed by many with an entrenched contempt.[81] Plant traders therefore endeavoured to avoid being besmirched with the connotations of baseness or selfishness that their occupation could convey.[82]

In both countries, they did so partly through what we might consider conventional means, such as visibly collaborating in scientific experiments with other respected scholars or publishing works on science. Reminiscing about the experimental planting schemes of Lord Petre, for example, merchant Peter Collinson noted that, 'I was an Eye Witness and Assistant in these Great Works'.[83] Through his research into the global distributions of heathers (*Erica* L.), James Lee 'discovered what islands had belonged to Europe and what to Asia', and his foreman John Cushing recalled how they had used an air pump to find out about the growth of plants.[84] But such activities were in fact only one aspect of the construction of scientific status. Practitioners also maintained their membership of the scientific community by adopting its mores and social comportment.[85] This held significant consequences for the ways in which breeders presented their work in public.

Scholarly research and communication was framed by the cultural milieu of the Republic of Letters, where the principles of politeness

---

[79] White, 'Purchase of knowledge', pp. 126–129.

[80] Hilaire-Pérez, 'Diderot's views', pp. 148–150; Scott, 'Authorship', p. 32.

[81] Chaussinand-Nogaret, *French Nobility*, pp. 84–90; Coquery, *L'hôtel aristocratique*, pp. 79–80. See also: Anon., 'Persons of quality', the author of which satirised the French nobility, who were, he claimed, obsessed with 'high birth . . . as if a Frenchman was only to be valued, like a black-pudding, for the goodness of his blood'. He contrasted this to the worthier British nobles, who 'not only often go into the city for a wife, but send their younger sons to a merchant's compting-house for education'.

[82] Crouzet, *Britain Ascendant*, p. 29.

[83] Marginal note by Collinson dated 30 October 1762, quoted in O'Neill, 'Peter Collinson's copies', p. 373. On eye witnessing and the construction of scientific facts, start with: Shapin and Schaffer, *Leviathan and the Air-Pump*, pp. 57–69, 336.

[84] Cushing, *Exotic Gardener* (1814), p. 29; Thornton, 'Sketch of the life of James Lee' (1810), p. xiv.

[85] Shapin, *Social History of Truth*, pp. 36–41.

often defined how members of this community of Enlightened individuals related to one another.[86] Joseph Banks perhaps epitomised this form of politeness. His carefully composed letters to his many correspondents show, for example, his assiduous care in smoothing over all disagreements, when they arose, in order to maintain harmony among members of the international scientific community.[87] The conventions of polite society required that members exercised self-control and modesty: most practitioners considered acts of unnecessary self-promotion, for example through publishing, in a very dubious light.[88]

These attitudes extended into the ways in which plants were named and claimed by their breeders. The tradition of giving honorific names to newly identified plant species was well established by the latter half of the eighteenth century, and formed part of the gift-giving culture that structured the social relationships among members of the botanical community (just as it did in the wider Republic of Letters).[89] This culture was completely distinct from that which determined the names given to plant varieties, which, as we have seen, were usually named after the person who claimed their 'discovery'. The tradition of granting honorific botanical names explains why neither Kennedy and Lee nor Andrieux and Vilmorin gave their own names to, or asserted their ownership of, the varieties they had bred.[90] The prevalence of these attitudes reinforced the ways that nurseries related to one another, and how they conceived of their productions.

Patents, Politeness and Plant Breeding

The 'polite' antipathy towards self-promotion was replicated in more general discussions about patenting. Throughout most of the eighteenth century, in both Britain and France, patents for any type of invention

---

[86] Gascoigne, *Joseph Banks*; Goldgar, *Impolite Learning*.

[87] De Beer, 'Preface' to Banks, *The Banks Letters*, p. x.

[88] These principles applied to all levels of the scholarly community, but they also varied according to national and institutional context. Banks himself published very little, yet nevertheless was elected President of the Royal Society in 1778. In contrast, the Académie des Sciences in Paris viewed publications as necessary to the construction of a scientific reputation. For more on this, see: Brockliss, *Calvet's Web*; Carter, *Sir Joseph Banks*, app. XIV, p. 571; Crosland, *Scientific Institutions*, 'Introduction'.

[89] For examples, see: Dr Henry Muhlenberg (Lancaster, Pennsylvania) to William Bartram (Pennsylvania?), 10 December 1792 and Richard Anthony Salisbury (Chapel Allerton, Yorkshire) to William Bartram (Pennsylvania?), 7 July 1793 in Bartram and Marshall, *Memorials*, pp. 472, 475. Muhlenberg wrote to Bartram about 'Cryptogamia' (mushrooms), and mentioned one that Linnaeus had recently named *Bartramia*. He declared that 'I love the little plant, now, twice as well, because it remembers me of such worthy friends.'

[90] Goldgar, *Tulipmania*, pp. 109–114.

were seen by many as being opposed to the ethos of polite society, in which acts of public spiritedness were very highly valued.[91] The notion that any 'polite' person would wish to claim intellectual property over a creation emerged only very gradually across the period. As Katie Scott has shown with regard to artistic and literary output, claims for patents and copyright were strongly associated with the marketplace. Men of letters consequently felt 'shame' in the idea that they might draw a personal advantage from, or claim exclusivity for, their productions.[92]

Similar attitudes extended into breeding plant varieties. The Royal Society of Arts, Commerce and Manufactures, for example, offered premiums and prizes to cultivators in Britain to encourage them to pursue horticultural or agricultural innovation. Improvements, for example, in the winter fodder available for animals or in extensive tree plantations were rewarded with cash prizes or honours such as medals. However, the new techniques and methods developed in response to this encouragement were emphatically *not* considered to be opportunities for personal profit. Patenting was anathema, as shown by the Society's *Rules and Orders* of 1765. 'No person', it asserted, 'will be admitted [as] a candidate for any premium . . . who has obtained a patent for the exclusive right of making or performing anything for which such premium is offered.'[93] In the early nineteenth century, Joseph Banks continued to envisage horticultural innovation within the same public-spirited discourse. On receipt of a gift of two new varieties of potato from the president of the London Horticultural Society, he proposed a communal experiment: Banks desired that new potato varieties should be distributed *gratis*, to 'every member [of the Horticultural Society] who is able to cultivate & report' upon them in the name of 'improvement'.[94] The insights gained from horticultural experimentation could then be extended to

[91] Berg, 'From imitation to invention', pp. 21–22; Goodwin, 'Thoughts on the question' (1786); Hilaire-Pérez, 'Diderot's views', pp. 146–147; MacLeod, *Inventing*, pp. 1, 186–187. On the changing context of inventing in eighteenth-century Europe, start with: Berg, *Luxury and Pleasure*, pp. 85–110; Berg and Bruland, *Technological Revolutions*, esp. pt. II.
[92] Scott, 'Authorship', pp. 29–30. On eighteenth-century notions of authorship, and on the debate about whether (and how) authors could claim intellectual property, start with: Hilaire-Pérez, 'Diderot's views', pp. 129–150; Scott, 'Art and industry', pp. 1–20.
[93] Wood, *History of the Royal Society of Arts*, pp. 115–116, 119ff, 145ff; quote re: patents p. 243. No patents were given for new breeds of plants, although several were issued for machines and techniques that aided their cultivation or processing. See: Woodcroft, *Subject-Matter Index*, vol. 1, pp. 19–30, 31–42, 360–361.
[94] NHM, DTC, vol. 18, f. 115, Joseph Banks (Soho Square, London) to Thomas Andrew Knight, 8 December 1811. See also: Anon., 'Observations on such nutritive Vegetables' (1783).

agricultural cultivation on a broader scale. Enlightened gardening might thus serve the general good.[95]

French agriculturalists similarly believed that the results of new improvements should be disseminated liberally throughout society. As already noted, Vilmorin gained widespread credit and respect for his public acts of altruism.[96] In the summer of 1788, for example, he philanthropically distributed seeds to farmers whose crops had been destroyed by a freak hailstorm in July. These were from his personal seed collection and were, specifically, varieties he had discovered that would grow either quickly or late in the season. Vilmorin's actions were key to his self-construction as a respectable and selfless scientific practitioner. They were also of significant advantage to his career.[97]

These methods of self-presentation paralleled those used by others working at the boundary between science and commerce. Bettina Dietz and Thomas Nutz have found that dealers in shells and other objects of natural history in Paris adopted the codes of behaviour of their customers, becoming '*curieux*' or 'amateurs' themselves.[98] Jonathan Barry and Michael Brown have shown that quack doctors and other medical practitioners in Britain also established their reputations through displays of literary pretension, the use of learned languages and, most especially, public acts of charity.[99] Like the upper-end plant traders, each of these groups operated in the interstices between science and the marketplace.

Breeders chose not to make proprietorial claims about their plants because they wished to associate their activities with those of the scholarly community rather than with the marketplace. Linnaeus had notoriously excluded the study of plant varieties from his definition of 'botany', and attacked the 'anthophiles', or lovers of flowers, who caused confusion by breeding new varieties. In his *Philosophia Botanica*, he asserted that, 'This introduction of varieties has dishonoured botany more than anything else', and elsewhere he exclaimed, 'I wish the system of varieties were entirely excluded from Botany & turned over entirely to the Anthophiles, since it causes nothing but ambiguities, errors, dead weight and vanity'.[100] Nurseries and plant breeders who cultivated a scholarly

---

[95] Jones, *Agricultural Enlightenment*, ch. 3.   [96] Silvestre, 'Notice biographique'.
[97] Vilmorin's benevolence served him particularly well during the Revolution, as will be discussed in Chapter 6.
[98] Dietz and Nutz, 'Collections curieuses', pp. 64–65.
[99] Barry, 'Publicity', pp. 29, 31, 33–34; Brown, *Performing Medicine*.
[100] Linnaeus, *Philosophia botanica*, Aphorism 259, translated by and cited in Stafleu, *Linnaeus*, p. 88; Linnaeus, *Hortus Cliffortianus*, cited in Stafleu, *Linnaeus*, p. 91.

reputation therefore had a strong interest in distancing themselves from the production of different varieties of plants.

In many cases, the rare plants that these nurseries sold had not yet been identified botanically.[101] Official names were awarded by established authorities, such as Carl Linnaeus or Philip Miller, but these were determined through consultation with the wider botanical community, which included plant traders. Linnaean botanist Jonas Dryander (1748–1810), who worked with Joseph Banks in London, paid no fewer than 659 visits to James Lee's nursery at Hammersmith between 1777 and 1798, where he examined the 'nondescript' (i.e. unidentified) plants that they grew.[102] Banks and his entourage also discussed plant species with other breeders: in one undated letter, the nurseryman James Dickson explained that 'I have sent you roots of the true Hermodaclys [sic] and Colchicum autumnale[.] It has been my opinion they are the same Plant.'[103]

It is noteworthy that Dickson did not attempt to decide *which* name the plant should have. Just as it was inappropriate to give one's own name to a plant, the unregulated attribution of any sort of botanical name was also severely frowned upon. In the Preface to his *Botanist's Repository* (1797), Henry Andrews decried the 'several changes of title' that many plants had undergone, and attacked the 'captious desire in every publisher, to foist something of his own coinage, upon the most trifling supposed difference'.[104] The immoderate naming of plants led to confusion and was a sign of an ungentlemanly inclination towards self-publicity. Botanically trained plant traders were therefore reluctant to accord names to plants independently, even when they had identified them.

As commercial businesses, nurseries competed to produce newer or better varieties of plants. But they played out this competition within the sphere of science, in which discussions of plant varieties rarely featured. Although there were undoubtedly rivalries between companies, and even though breeders did compete for the same pool of consumers, they were not out-and-out adversaries. Some nurseries collaborated, sharing

[101] For examples, see: BL, Ms. 8097, f. 192, L'Héritier de Brutelle (Paris) to Joseph Banks (London), 28 May 1789; BNF, NAF 2758, ff. 111–114, Tschüdy (Colombé) to Vilmorin-Andrieux (Paris), 7 May 1813; BNF, NAF 2758, f. 122, Vilmorin (Paris) to [unidentified recipient].

[102] Carter, *Sir Joseph Banks*, app. XII. Dryander visited the nursery more than once a fortnight for twenty-one years.

[103] NHM, Botany Library, Banks Correspondence, f. 148, Jas: Dickson (Covent Garden) to Sir Joseph Banks, Thursday, 12 o'clock.

[104] Andrews, *Botanist's Repository* (1797), Preface.

information about new techniques for cultivating plants and about new technologies for housing and caring for plants. Nurseries would exchange specimens with one another, sometimes as a straight sale, but also as swaps, in order to improve general knowledge about plant cultivation and to share expertise.[105] The affiliations within the nursery network are very evident in the court trial quoted earlier, in which plaintiffs and defendants called witnesses and alibis from the wider community. Nurseryman Daniel Grimwood (plaintiff) explained that 'one of those plants so missed, was one that we has [sic] bought of Mr. Colvill, a nursery man in Chelsea'. Called to the witness box, James Colvill asserted that he knew Grimwood 'very well', and that he had helped him to search the house of the accused.[106] Grimwood and Colvill may have competed for customers, but they also socialised with, and supported, each other.[107]

There were three layers to the plant traders' public self-presentation in the eighteenth century. At heart, they were merchants who sold goods within an expanding marketplace. But the more elite traders acted as intermediaries who connected botanists and gardeners of all types. They were important because they translated and negotiated the demands and desires of both groups, and because they made possible the wider cultivation of rare ornamental species.[108] Although the upper-end traders did not seek gentlemanly status, they presented themselves in ways akin to the polite 'gentlemen' or 'amateurs' of science, for example by gifting plants to other scholars. Conforming to the ideals of polite science, they considered that the exchange of specimens in this way was a means to gain social promotion via patronage networks.[109] Finally, these elite traders deepened the association with polite science through their philanthropy, which was well received by other scholars and by the general public. Their munificence enhanced their ability to obtain new plants

---

[105] Cushing, *Exotic Gardener* (1814), pp. viii–ix; Willson, *James Lee*, p. 31.

[106] OBP, 16 September 1795, trial of Charles Fairfield (t17950916–73).

[107] Sometimes, this socialisation could be to the detriment of science, rather than to its gain. In 1786, Joseph Banks described to one correspondent how 'The Nat: Hist: Society you enquire after does not seem to gain ground[.] Lee of Hammersmith is a leading man & he I hear sometimes . . . adjoins the meeting at the Black Bear in Piccadilly where they drink grog 'til they do not know monandria from Cryptogamia'. NHM, Botany Library, Banks Correspondence, f. 145r, Joseph Banks (Soho Square, London) to Marmaduke Tunstall (Wycliffe, Yorkshire), 19 February 1786. Monandria and Cryptogamia were two Linnaean classes of plants. Emphasis in original.

[108] Miller, *Gardener's Dictionary* (1752), p. 3.   [109] Spary, *Utopia's Garden*, p. 63.

from other scholars and traders. The elite nurserymen's private interests were thus balanced by their claim that they held the public good at heart. The socio-cultural conventions of eighteenth-century science demanded that the commercial rivalry between these traders must be muted.

3   Amateur Botany

Writing about James Lee in 1810, the physician and writer on botany Robert John Thornton divided the Vineyard Nursery's customers into two groups. First, there were 'lovers of botany', who were motivated by 'a laudable enthusiasm in the pursuit of knowledge'. These were people, Thornton claimed, to whom Lee would 'give duplicates away' as gifts rather than sell them as commodities.[1] In other words, they were botanically competent consumers who aimed to create specimen collections of their own. In addition, however, Lee dealt with a second cohort of consumers, whom Thornton dismissed as 'rich but careless collectors of flowers'. This latter group, he declared, was 'led to them [rare plants] through ostentation'.[2] The implication was that these collectors were concerned with luxury, superficiality and self-display, and not with the commendable search for new knowledge. While the former selected specimens carefully, using specific 'scholarly' (botanical) criteria, the latter were satisfied with any plant that gave them aesthetic pleasure.

Thornton's portrayal of public participation in plant collecting placed the intellectual and the aesthetic in opposition to each other. However, the division of the public into the unassumingly knowledgeable and the ostentatiously ignorant obscures huge diversity in levels of understanding, in the range of practices undertaken and in the values attributed to different kinds of enquiry. For most of the eighteenth century, the 'amateur' scholar was expected to cultivate a refined aesthetic sensibility (largely through the appreciation of art), as well as a respectable understanding of science. Negative judgements cast upon amateur botanists were influenced more by assumptions about the gender and social status of the practitioner than by concern about 'aesthetic' approaches to the science *per se*. Art and science were thus not nearly as incompatible as Thornton's characterisation might suggest.

This chapter discusses the public with whom the commercial nurseries interacted. We saw earlier that the nurseries marketed their exotic

---

[1] Thornton, 'Sketch of the life of James Lee' (1810), pp. xv–xvi.   [2] Ibid.

stock as specimens of interest to consumers identified as 'scholars'. What did that identification with scholarship mean in terms of practice and participation, however? The discourses about amateur scholarship that circulated in the late eighteenth and early nineteenth centuries were understood and interpreted in multiple ways by the different people interested in studying botany. Public participation in science was also guided – and delimited – by notions of propriety and education that were gendered and classed. Contemporary formulations of the 'scholar' placed great emphasis on connoisseurship and the cultivation of taste, for example – attributes that were socially exclusive by their very definition. Yet, middle-ranking men, and middling- and upper-ranking women, devised means of negotiating the dominant association between scholarship and elite masculinity. In investigating the articulation of amateur scholarship, this chapter uncovers multiple cultures of botany that emerged through the mediation of those associations. It offers a portrait of a culture and society within which commercial nurseries emerged as botanical intermediaries. The values placed upon alternative forms of amateur botany were consistently bound up in notions about propriety, gender and social status. These alternative botanical cultures were often orientated towards practical applications of botanical knowledge that, although culturally associated with the lower classes or with women (of any rank), ultimately came to be valued as the new science of horticulture.

Scholarly Culture in Discourse

Robert John Thornton's bisection of James Lee's customers repeated a distinction that had become a commonplace by the early nineteenth century. Sixty years previously, the two great eighteenth-century authorities on natural history and botany, Georges-Louis Leclerc, Comte de Buffon, and Carl Linnaeus, had articulated the same sentiments. Buffon noted disparagingly that 'most of those who, without any prior knowledge of natural history wish to have collections... are people of leisure with little to occupy their time otherwise, who are looking for amusement, and regard being placed in the ranks of the curious as an achievement.'[3] Linnaeus made a similar differentiation among students of nature, attacking the 'anthophiles', or lovers of flowers, whose interest in plants' aesthetic appearance was mere trivia compared to the valued observations made

---

[3] Buffon, 'Initial discourse'. Note that this opening discussion was not included in the eighteenth-century English translations of Buffon's *Histoire Naturelle* (by Kenrick & Murdoch, Smellie and Barr), so Buffon's sentiments may not have been widely known across the Channel.

by real botanists.[4] A line had been drawn among the various publics that might practise natural history. On one side stood a select few, praised by Buffon, Linnaeus and their disciples for their scientific understanding of the natural world. On the other side resided an ostentatious majority, whose appreciation of nature appeared merely aesthetic and inconsequential.

The public served by the upper-end nurseries are perhaps better conceived of as three, rather than two groupings. Plant traders interacted with botanical 'authorities' (such as Linnaeus and employees of the various botanical gardens); aristocrats and the wealthiest members of the upper ranks; and finally a more amorphous 'wider public', largely comprising members of the middling ranks. Some of the wealthiest members of the latter two groups purchased unusual ornamental plants simply for inclusion in their gardens. Others, including those of lesser means, bought rare plants for study in their personal collections. Members of each of these groups might consider themselves to be amateurs and/or scholars.

It is challenging to describe and delineate the diversity within eighteenth-century scholarly culture. The absence of professional qualifications meant that neither the French nor the English language possessed a clear terminology with which to articulate the fine distinctions among levels of expertise. The word 'curious' was often used generically to denote either someone who engaged in learning or an object worthy of study. A 'curious' collector might also be called (for example) a connoisseur, a virtuoso,[5] a dilettante, an antiquary, a pedant, a savant, a Bluestocking or a *précieuse*.[6] In his recent study of Britain's Society of Dilettanti, Jason Kelly has discussed how the word 'dilettante' existed 'at the nexus of competing discourses, meanings, and practice about social status and gender'.[7] This blurred meaning was not specific to dilettanti.

[4] Stafleu, *Linnaeus*, pp. 90–91.

[5] Somewhat confusingly, 'virtuoso' might refer to someone engaged in a narrow, self-interested study or to the exact opposite: someone with interests that were excessively wide-ranging. In contrast, proponents of connoisseurship presented the connoisseur as someone who achieved a balance between specificity and generality. See: Bermingham, 'Elegant females', p. 504; Gascoigne, *Joseph Banks*, ch. 3, esp. pp. 57–66. On the transition in meaning of 'virtuoso' over the seventeenth and eighteenth centuries, start with: Jackson Williams, 'Training the virtuoso', pp. 157–182; Pace, 'Virtuoso to connoisseur', pp. 166–188; Yeo, *Notebooks*, Introduction. By the 1780s in France, 'virtuoso' could also denote someone who demonstrated great technical skill in an art or craft, and who performed their talents in front of an audience. See: Metzner, *Cresecendo of the Virtuoso*, p. 1.

[6] Bending, 'Every man', pp. 520–521; Bhattacharya, 'Family jewels', p. 212; Kelly, *Society of Dilettanti*; Pomian, *Collectionneurs*; Redford, *Dilettanti*; Shapin, 'A scholar and a gentleman', pp. 279–327 and 'The image of the man of science'.

[7] Kelly, *Society of Dilettanti*, p. 8.

The labels used to describe all eighteenth-century scholars were seman-tically capacious, subject to discursive interpretations that were variously positive or negative and that invoked competing ideas about the gender and social status of the scholar. The fluidity in meaning left open a broad landscape for a range of engagements with scholarship.

The epithets of 'botanist' and 'naturalist' were normally reserved for people identified as botanical authorities – the small number who worked at the botanic gardens, for example, who were professionally occupied in the scientific study of plants. Nurseries and writers on botany would instead describe people who collected and studied plants as 'amateurs' or 'lovers'. These terms remained largely complimentary until well into the nineteenth century. In 1817, for example, 'a Deputation of the Cale-donian Horticultural Society' travelled from Scotland to the Low Coun-tries and northern France to assess 'the state of Horticulture' there. In the account of the expedition, published in 1823, the authors stated that their primary aim was to visit the 'principal amateur cultivators and pro-fessional nurserymen', with whom they hoped 'to establish a correspon-dence' and from whom they hoped to obtain 'new or uncommon vari-eties of fruits and culinary vegetables'.[8] It is worth underlining here that the term 'amateur' was used consistently to refer to someone who did not make a commercial profit from their plant collections, and the word's eighteenth- and early nineteenth-century meaning otherwise contrasts with that of the present day. While the term now often implies a defi-cient engagement with knowledge, the Caledonian Horticultural Society and their readers normally used 'amateur' as a respectful referent to non-professional collectors of plants.

The predominant discursive construction of amateur scholarship in late eighteenth- and early nineteenth-century Britain and France was an approving one. The *Oxford English Dictionary* dates the first recorded use of 'amateur' in English to 1784, although it is likely that the term was used in the 1770s (if not earlier) as a French import. The eighteenth-century English and French meanings of 'amateur' were close to the original Latin 'amator', signifying that a person 'loved' the sub-ject of study.[9] Associated with the figure of the connoisseur, amateur

[8] Neill, *Journal of a Horticultural Tour*, p. vii.
[9] 'Amateur' in OED Online, http://www.oed.com/view/Entry/6041 [accessed 1 February 2016]; Imbs, *Trésor*, vol. 2, pp. 675–681. Note that the English translation of Krzysztof Pomian's book *Collectionneurs, amateurs et curieux* uses the word 'dilettante' for what in French is termed 'amateur'. This is appropriate for the context of fine art collecting in Britain, but it was very rare to find naturalists describing themselves as 'dilettanti'. Instead, they were more likely to refer to themselves as 'lovers of nature' or 'lovers of botany'. I have therefore chosen to retain the term 'amateur', because it conveys the same sense of being enamoured with nature. Compare Pomian, *Collectors and Curiosities*, pp. 132–138 and *Collectionneurs*, pp. 154–162.

scholarship was firmly situated within a conception of polite scholarly culture in which individuals were encouraged to develop a broad purview through studying a variety of subjects, ranging from the fine arts to the sciences. To be an amateur, therefore, ideally involved more than simply studying botany.

While the precise meanings of the terms 'amateur' and 'connoisseur' remained fluid (especially with regard to the study of nature), French writers made a hierarchical distinction between the two formulations. The *Encyclopédiste* Paul Landois discussed connoisseurship in relation to the fine arts, and emphasised that a connoisseur 'is not the same thing as an *amateur*'.[10] The theory went that, while both connoisseurs and amateurs publicly displayed their taste and skill through their collections, connoisseurs were pre-eminent because they actually practised their subject of study. 'One is never a perfect connoisseur of Painting', Landois declared, 'without being a Painter'.[11] Practical knowledge would increase one's understanding of the theory relevant to the topic, and would lead to more accurate aesthetic judgements. By contrast, amateurs of art felt a passion – 'love' – for their subject, but were not practitioners.[12] The figure of the amateur nevertheless retained positive cultural associations more broadly: we saw earlier that the French plant traders flatteringly referred to the readers of their catalogues and advertisements as 'amateurs'. British writers (who were less concerned with parsing out semantic distinctions) used the terms 'amateur' and 'connoisseur' more or less interchangeably, as complimentary descriptions of someone engaged in learning. The notion of being an amateur or connoisseur was culturally attractive to members of the upper-middling ranks and above, and the concepts associated with it underpinned much botanical collecting.

Connoisseurship gained in cultural value because it was increasingly considered central to the identity of the gentleman. In the sixteenth and seventeenth centuries, the figures of the 'scholar' and the 'gentleman' had been set in opposition to each other. Scholars were considered to be withdrawn from society, contemplating abstract problems in solitude.[13] In contrast, a gentleman aimed to fulfil his civic duty by taking an active

---

[10] Landois, 'Connoisseur' (1753); Pomian, *Collectionneurs*, pp. 155–156. The amateur Jean-Georges Wille repeated this phrase from the *Encyclopédie* almost word for word in 1776, in his description of the baron of Buchwald: 'he is an amateur, but every amateur is not a connoisseur.' Guichard, *Les amateurs*, p. 166.

[11] Landois, 'Connoisseur' (1753).

[12] According to the *Encyclopédie*, an amateur 'loves art and has a decided taste for paintings'. Landois, 'Amateur' (1751). See also: Dietz and Nutz, 'Collections curieuses', p. 54; Pomian, *Collectionneurs*, p. 158.

[13] Shapin, 'A scholar and a gentleman', pp. 282, 289 and 'The mind is its own place', pp. 198–200.

engagement in society.[14] The conceptual opposition between the character, culture and comportment of the scholar and the gentleman had largely disappeared by the early eighteenth century, thanks to the rise of the 'new science' of Francis Bacon and Robert Boyle.[15] The practice of science was now situated within the same codes of comportment as those followed by a gentleman, and gentlemen, therefore, were increasingly expected to participate in the kind of scholarly practice associated with the figure of the connoisseur.[16]

Contemporary writers emphasised that the connoisseur should unite the civic humanist ideals of disinterestedness and service to the nation with 'gentlemanly' impartiality and private virtue. Jonathan Richardson explained in 1719 that

To be a connoisseur ... a man must be as free from all kinds of prejudice as possible; he must moreover have a clear and exact way of thinking and reasoning; he must know how to take in, and manage just ideas; and throughout he must have not only a solid, but unbiased judgement.[17]

In behaving like a gentleman, the eighteenth-century connoisseur showed that he could be relied upon to uphold the 'truth'. This was essential to the communication of information, and thus to the construction of new knowledge.

By the mid eighteenth century, the study of science was promoted in Britain and France as an activity beneficial both to the individual and to society. If practised properly, scientific learning would lead to positive cultural and social outcomes, forming a patrician class with a morally upstanding character.[18] Anthony Ashley Cooper, third earl of Shaftesbury (1671–1713), one of the most influential writers on polite science, explained that,

I am persuaded that to be a *Virtuoso* (so far as befits a gentleman) is a higher step towards the becoming a Man of Virtue and good Sense, than the being what in this age we call *a Scholar*. For even rude Nature it self, in its primitive Simplicity, is a better Guide to Judgement, than improv'd Sophistry, and pedantick Learning ... The mere *Amusements* of Gentlemen are found more improving than the profound *Researches* of Pedants.[19]

---

[14] Shapin, 'A scholar and a gentleman', p. 289.   [15] Ibid., pp. 292–298.

[16] Gascoigne, *Joseph Banks*, pp. 58–60; Klein, *Shaftesbury and the Culture of Politeness*, pp. 34–35, 41; Nye, 'On gentlemen', p. 328.

[17] Jonathan Richardson, 'The science of a connoisseur' (1719, repr. 1773), quoted in Bermingham, 'Elegant females', p. 504.

[18] On science and sociability, see: Belhoste, *Paris Savant*, ch. 5; Golinski, *Science as Public Culture*; Schaffer, 'The consuming flame'; Stewart, *Rise of Public Science*.

[19] Shaftesbury, *Soliloquy* (1710), pp. 174–175. See also: Kelly, *Society of Dilettanti*, p. 21.

According to Shaftesbury, a refined, sociable education would counter-balance (and prevent) the negative effects of excessive study and would foster rational social civility. Like other contemporary writers on polite-ness and education, he also insisted that his pupils cultivate a sense of moderation and balance in all things.[20] Scholars were consequently very attentive to how their behaviour might be perceived. In 1752, for exam-ple, Emanuel Mendes da Costa (1717–1791), a London-based vendor of shells and minerals, acknowledged apologetically in a letter that 'my desire of learning makes me surpass the bounds of Decency... I rely intirely [sic] on your Candour & humanity to solve my demands'.[21] Da Costa was concerned that his own thirst for knowledge might seem to be immodestly unrestrained, and therefore distasteful to his correspondent.

Shaftesbury differentiated the gentlemanly 'virtuoso' from the 'scholar' and from 'pedants'. Too much learning was considered detri-mental to good character, leading to narrowness of view and encouraging obsessive behaviours.[22] Indeed, conduct writers were unanimous in con-demning the excessive specialisation and unrestrained displays of knowl-edge associated with the pedantic scholar. 'From the moment you are dressed and go out,' the Earl of Chesterfield told his son in the 1740s, 'pocket all your knowledge with your watch, and never pull it out in company unless desired.'[23] A gentleman was not to appear too learned in public.

In Britain, the antipathy towards public displays of knowledge extended to publishing. In his *Introduction to Botany*, James Lee thanked an anonymous 'Gentleman' who gave him assistance. He explained that he 'would willingly have informed the Public to whom he was obliged, had he not been prohibited, while that Gentleman was living, from men-tioning his name'.[24] Modesty was also the reason given for why nurs-eryman James Gordon did not write for publication.[25] The strictures regarding public displays of knowledge were as much about class as gender.

---

[20] Klein, *Shaftesbury and the Culture of Politeness*, 'Introduction' and *Shaftesbury: Charac-teristics*.

[21] WCRO, CR 2017/TP408, f. 2/b/r, Draft letter, Emanuel Mendes da Costa (London) to Thomas Pennant (Downing), 4 April 1752.

[22] Kelly, *Society of Dilettanti*, pp. 21–24. See also: Levine, *Battle of the Books*.

[23] Philip Dormer Stanhope, Earl of Chesterfield, *Letters to his Son and Others* (c. 1746–53), quoted in Shapin, 'A scholar and a gentleman', p. 303.

[24] Lee, *Introduction to Botany* (1765), p. xv. Adrian Johns discusses how gentlemen scholars in Britain had to negotiate two contradictory demands: on the one hand, a gentleman shunned authorship; on the other, an author's claim to authority rested on their status as a gentleman. Johns, 'Natural history as print culture', pp. 110–111.

[25] Grieve, *Transatlantic Gardening Friendship*, p. 22.

Attitudes towards publishing were markedly different in France, where, by the second half of the eighteenth century, publishing was deemed essential for establishing academic merit, and for securing and reinforcing patronage.[26] France's provincial academies had run prize competitions regularly since the seventeenth century, to which the public was invited to submit poems or essays on specific topics. The competitions promoted publication as a goal for many French scholars – including women.[27] The Académie des Sciences likewise judged academic excellence on the basis of publications, granting its approbations only to works deemed of sufficient quality. The Académie's publications committee effectively set standards for scholarship across France.[28] The contrast between French and British attitudes towards publishing complicated the construction of reputations in a transnational context, however. One of the stated reasons why the Académie would not elect British naturalist Joseph Banks to the status of correspondent in 1783 was that he had published too little.[29]

Amateur botany was thus firmly situated within the culture of polite, gentlemanly science. The introduction to Conrad Loddiges & Sons' *Botanical Cabinet* (1817) crystallised its characteristic features. 'Our aim', Loddiges explained, 'is to direct the minds of those who may honour us with their patronage, to a source of amusement at once intellectual, exalted, delightful and unbounded.'[30] He continued:

[Botany] is assuredly no dry or abstruse study; it is a perpetual spring of the most genuine satisfaction. Even when cares or troubles assail the mind, and overshade all things with gloom (and no one is always exempted from such things) even then let us look at these beauties – let us contemplate them. Yes! We will 'consider the Lilies how they grow;' our Divine Saviour himself commands us to do it.[31]

Conrad Loddiges framed amateur botany as an 'amusement' that was an antidote to the 'cares' of everyday life. Loddiges was also at pains to emphasise that although the study of nature offered more than trifling pleasure, it was never 'dry or abstruse', and would not lead to pedantry. Botany was 'at once intellectual, exalted, delightful and unbounded', and it was morally virtuous, ultimately conducting its students towards divine appreciation.

[26] Brockliss, *Calvet's Web*, pp. 85–87.
[27] Caradonna, *Enlightenment in Practice*, pp. 32–39.
[28] Hahn, *Anatomy*, ch. 3, esp. pp. 60–63.
[29] NHM, Botany Library, DTC, Vol. 3, ff. 49–53, Charles Blagden (Paris) to Joseph Banks, 18 June 1783. Banks was made a foreign member in 1787. Miller, 'Joseph Banks', p. 21.
[30] Loddiges and Sons, *Botanical Cabinet* (1817), Introduction [unpaginated].   [31] Ibid.

Women as Amateurs

The formulation of polite science as 'gentlemanly' contains a clear gender bias, but in spite of this discursive association women were excluded neither from learning nor from the study of botany. Most eighteenth-century pedagogical writers considered that some scholarship was appropriate for the female mind, although most also concurred that the ultimate aim in educating a middle- or upper-ranking girl was to fit her for her civic role as a wife, a mother and the mistress of a house.[32] Jean-Jacques Rousseau put this most starkly in *Émile*, where he declared that 'all women's education must be related to men', and his sentiments were widely replicated by male and female writers on both sides of the Channel.[33]

Science was not excluded from women's curricula, however. Contemporary writers on female education were united in their dislike of the excessively erudite *'femme savante'*, but late eighteenth-century pedagogical writers such as Mary Wollstonecraft (1759–1797) and Marie Le Masson Le Golft (1749–1826) nevertheless recommended that women should be equipped with an up-to-date practical knowledge of science.[34] Subjects such as botany and chemistry were particularly recommended. As the educational writer and novelist Maria Edgeworth (1768–1849) explained in 1798, botanical and chemical knowledge could have a practical application within the home and could be taught safe in the assurance that 'there is no danger of inflaming the imagination.'[35] Writers also recommended that women should take up gardening in conjunction with the study of rational scientific subjects.[36] Wollstonecraft suggested that 'women in the middle rank of life' should devote their time to managing

their families, instruct[ing] their children, and exercis[ing] their own minds. Gardening, experimental philosophy [i.e. natural science], and literature, would afford them subjects to think of and matter for conversation, that in some degree would exercise their understandings.[37]

The content and purpose of female scientific education mirrored that devised for polite 'gentlemen'. Writers on both male and female education shared a horror of excessive individual learning and of ostentatious

---

[32] Bolufer Peruga, 'Introduction', p. 192.
[33] Rousseau, *Émile* (1762), quoted in Bolufer Peruga, 'Introduction', p. 191. See also: More, *Strictures* (1799); Wakefield, *Reflections* (1798), ch. III.
[34] Bloch, 'Discourses of female education', p. 248.
[35] Benjamin, 'Elbow room', pp. 35–36; Edgeworth, *Letters for Literary Ladies* (1798), pp. 20–21, quote p. 20; Kelly, *Women, Writing and Revolution*, pp. 11–12.
[36] Wakefield, *Reflections* (1798), pp. 91–92, 136–138.
[37] Wollstonecraft, *Vindication of the Rights of Woman* (1792), p. 195.

displays of knowledge. For women as much as for men, knowing some science was beneficial to individual development and to society at large – although women who deviated from the norm (in this case by knowing too much) were subject to harsher censure.

Although some scientific knowledge was recommended for women, very few described themselves as scholars, amateurs or connoisseurs. These epithets were conventionally reserved for men – and as such were rarely considered complimentary for women. As explained by 'Hortensia' in Maria Elizabeth Jacson's *Botanical Dialogues* (1797), 'information in a woman, beyond a certain degree, distinguishes her above her companions, and . . . is liable to bring her into a vain display . . . Hence she becomes ridiculous.'[38] Contemporary criticisms of female scholars closely mirrored those of excessively scholarly gentlemen: in Britain, the learned woman was castigated as a 'female pedant'; in France, she was a *précieuse*, or a *femme savante*.[39] Injurious stereotypes of female connoisseurs depicted them as impolite, 'dilettante' and immodest.[40]

Such unflattering formulations did not mean that all women eschewed scholarship, however. Female participation in eighteenth-century scholarly culture can be divided broadly into two camps: a minority of women emulated aspects of male scholarly culture, while a majority adapted it in ways consonant with contemporary notions of female propriety. As we will see later, some members of the former group were to gain recognition as connoisseurs or amateurs, but risked social censure in doing so. The latter developed forms of intellectual engagement which were less socially problematic, but often far from the norms of (gentlemanly) polite science.[41]

What kinds of woman, then, might emulate amateur scholarship? A few, such as Margaret Cavendish Bentinck (1715–1785), duchess of Portland, publicly adopted connoisseur-like behaviour, forming collections, exchanging specimens, acting as patrons and corresponding with other (male) scholars. By the time of her death in 1785, Portland's collection of natural history and art was reputed to be the second largest in Britain (the largest was that of Hans Sloane). Portland taught botany to Jean-Jacques Rousseau while in England in 1766–67, and also offered financial and social support to a glittering array of botanists,

---

[38] Jacson, *Botanical Dialogues* (1797), p. 239; Shteir, 'Botanical dialogues', p. 315.

[39] Bolufer Peruga, 'Introduction', p. 193; George, 'Linnaeus in letters', p. 13; Outram, *Enlightenment*, pp. 91–92.

[40] Shteir, 'With matchless Newton', pp. 120–121, 127–128.

[41] My approach here has been influenced by Linda Nochlin, 'Women artists' and by the debates and discussions that ensued from the publication of this essay.

including the botanical artist Georg Dionysius Ehret, Philip Miller (head gardener of Chelsea Physic Garden) and the Linnaeans Daniel Solander and Joseph Banks.[42] Women like Portland who openly participated in eighteenth-century scholarly culture were exceptional, however. The female amateurs whose scholarship was made public mostly occupied one of two ends of a spectrum, either being relatively independent thanks to the possession of significant financial resources and substantive social capital, or being forced into publishing by financial necessity.[43]

It was conventional for most women who emulated the model of amateur scholarship to study and collect in private – the scant traces that such activities leave in the archives makes the task of recovering their work somewhat challenging. One means of finding out about female amateur scholars, however, is to study those who ultimately did gain wider recognition for their work. The naturalist Anna Blackburne (c. 1726–1793) offers a good example. Blackburne formed a huge zoological and botanical collection in her ancestral home at Orford Hall, near Warrington, where she lived unmarried with her father John Blackburne (c. 1694–1786).[44] Unlike the Duchess of Portland, she took care to remain out of the public eye for most of her life.

It is not known when Blackburne began collecting and studying natural history, but she probably started from a young age. Natural history collecting was a shared family passion, and Blackburne initially obtained her specimens via her father, cousin and brothers, who all collected, corresponded and travelled extensively.[45] Her early intellectual activities were further aided by the proximity of a non-conformist

---

[42] George, *Botany*, p. 5, 9; Hayden, *Mrs Delany*, pp. 105–111; Rogers, 'Bentinck, Margaret Cavendish'.

[43] Shteir, *Cultivating Women*, ch. 2.

[44] Anna Blackburne's huge natural history collection consisted of a herbarium, insect and marine specimens, minerals, ores and fossils, as well as a taxidermy collection comprising 470 birds and one bat. WLA, RWA Wp 82182, 'Orford Hall List of Natural History Specimens, Portraits etc' (Warrington Museum and Art Gallery, 1914). Blackburne moved to the newly built Fairfield House after her father's death, where her museum occupied a room of fifteen yards length, which spanned the entire width of the house. *Gentleman's Magazine* 94 (1824), p. 210, cited in Wystrach, 'Anna Blackburne', p. 162. See also: TNA, Prob 11/1241, Will of Anna Blackburne; Carter, *Warrington*, p. 56; Shteir, 'Blackburne, Anna'.

[45] Wystrach, 'Ashton Blackburne's place', pp. 607–608 and 'Anna Blackburne', pp. 151, 161. Blackburne's cousin was Ashton Lever, whose 'Leverian Museum' comprised a gigantic collection of natural history, antiquities and art. Lever was one of the major recipients of specimens brought back on Captain Cook's voyages. The museum was moved from Manchester to London in 1775, where, due to its size and comprehensiveness, it soon became an essential port of call for connoisseurs in London. Torrens, 'Natural history', p. 83.

college, the Warrington Academy.[46] The Academy was notably progressive with regards to women's education: it did not officially admit female students, but women were nevertheless permitted to attend classes. Anna Laetitia Aikin (later Barbauld), daughter of theology master John Aikin, recalled how 'a fine knot of lassies' studied with her at the Academy in the 1760s.[47] While there is no evidence to confirm whether or not Blackburne herself studied at the Academy, she and her father maintained close links with several masters and their families, socialising with them at Orford Hall. She may also have been acquainted with this 'knot' of intellectual Dissenting women.

The extant sources show that Blackburne became known to scholars outside her family and the local community by the late 1760s. At this point in her life, however, she was in her forties, established as mistress of Orford Hall and firmly off the marriage scene. Her age and respectable position may have offset the less agreeable connotations of female scholarship. Blackburne formed a strong intellectual friendship with the naturalist Johann Reinhold Forster (1729–1798), who taught at the Academy between 1767 and 1770.[48] In 1768, Forster described how he walked to Orford every Saturday, where 'I dine ... & range & make out the Insects of Miss Blackburne & read to her my lectures on Entomology'.[49] In return, Blackburne gave him access to the extensive family library.[50] When she wrote to Carl Linnaeus in 1771, she discovered that her (complimentary) intellectual reputation had preceded her: in his reply, the Swedish botanist recalled hearing about a trip Blackburne had made to Oxford two years previously; he enquired whether

---

[46] On the Warrington Academy and its connections with science, see: Fulton, 'The Warrington Academy'. For the links between the Academy and non-Dissenters, see: Wakefield, *Memoirs* (1792), p. 199.

[47] Uglow, *The Lunar Men*, pp. 73–75, quote p. 73.

[48] Forster sailed as the naturalist on Captain Cook's second circumnavigation. He named a newly discovered New Holland genus of plants *Blackburnia* in honour of Anna and John Blackburne (the plant has since been placed in the genus *Zanthoxylum* as *Z. blackburnia*). Carter, *Warrington*, p. 99; Wystrach, 'Anna Blackburne', pp. 157–158. For the garden at Orford, see: Neal, *Catalogue* (1779).

[49] J. R. Forster to Thomas Pennant, 30 November 1768, cited in Wystrach, 'Anna Blackburne', p. 157.

[50] The relationship was close enough for Forster to expect Blackburne to order books of interest to him, and it was a source of frustration to him when she did not do so. In 1769, he wrote to Thomas Pennant, exasperatedly exclaiming that 'Miss Blackburne the Daughter of the rich Esq. ... made some trifling acquisitions in books & told me that Rosel [a book he wished her to purchase] was too dear for her. If she says thus, what must other people do?' J. R. Forster to Thomas Pennant, 19 June 1769, cited in Wystrach, 'Anna Blackburne', p. 158. The book mentioned by Forster was probably Rosel's *History of Frogs*, which Pennant had purchased in Germany for the connoisseur William Constable in 1765. See: Bodleian, Ms. Eng. Let. c.229/159r, Thomas Pennant (Downing) to William Constable, 6 October 1765.

she was one of 'three botanical ladies ... who in 1769 disputed with and triumphed over the botanist in the [botanic] garden at Oxford about the geranium'.[51] Linnaeus later sent one of his students to examine her insect collection.[52]

Blackburne's initial connections led to further links with other naturalists. Following his move from Gibraltar to Lancashire in 1773, the naturalist and vicar John White became acquainted with her.[53] White probably helped to engineer an introduction to his London-based brother, the publisher Benjamin White, through whom Blackburne then exchanged specimens with Peter Simon Pallas in Russia.[54] In 1774, Blackburne received a surprise letter from the now impoverished naturalist Emanuel Mendes da Costa, asking whether she might employ him to catalogue her mineral collection. She was initially keen to receive him, although his tarnished reputation (due to his having embezzled around £1500 of the Royal Society's funds) seems to have prevented her from ultimately accepting the offer.[55] Blackburne also made the acquaintance of Joseph Banks and Daniel Solander, just as they were rising to scholarly eminence in the mid 1770s.[56]

Meeting and corresponding with male scholars constituted a relatively limited level of publicity. Publicity through print was a quite different matter, and the social condemnation heaped on women who ostentatiously published was generally more severe than that directed towards men.[57] Anna Blackburne did not publish anything herself, and when

---

[51] Carl Linnaeus (Upsala) to Anna Blackburne (Orford), 28 July 1771. Letter translated from the original Latin and reproduced in Beamont, *History and House of Orford*, p. 62. Ann B. Shteir notes that Linnaeus was mistaken about the 'three botanical ladies': Anna Blackburne was in fact the only woman present. Shteir, *Cultivating Women*, pp. 53–54.

[52] Wystrach, 'Anna Blackburne', p. 155.

[53] John White had supplied his brother Gilbert with numerous bird specimens from Gibraltar and observations about the migration of birds. White, *Natural History* (1789), pp. 72–73, 191, 209–210. He also corresponded with Linnaeus between 1771 and 1774.

[54] Wystrach, 'Anna Blackburne', p. 155.

[55] BL, Add. Ms. 28534, f. 282v, Draft letter, Emanuel Mendes da Costa (London) to Anna Blackburne (Orford Hall), 30 August 1774; BL, Add. Ms. 28534, f. 233, Anna Blackburne (Orford Hall) to Emanuel Mendes da Costa (London), 12 October 1774. See also: Foote, 'Mendes da Costa, Emanuel'.

[56] In a letter to Linnaeus of 1771 (quoted in Wystrach, 'Anna Blackburne', p. 154), Blackburne mentioned that she had not yet made the acquaintance of Banks and Solander, who had recently returned from their voyage on the *Endeavour*. This introduction had been achieved by Spring 1779, when Pallas wrote to Banks and asked him to tell 'Mrs. Blackburne at Orford, of a Parcell [sic] of seeds & a Box with ores being sent for her ... & that I have not yet received any of the articles she promised to send me early in Spring'. BL, Add. Ms. 8094, f. 240r, P. S. Pallas (St Petersburg) to Joseph Banks, 18/29 May 1779.

[57] The few women who *did* publish on botany consistently prefaced their work with apologies, asking their readers to excuse them for placing their work in the public domain.

her name did appear in print she was always presented as a helpmeet to another male scholar, where her role was decidedly passive.[58] Published descriptions of Orford Hall and its collections thus focused more on Anna's father and brothers than on herself, even though she was responsible for the major part of the collections.[59] Thomas Pennant described 158 of her zoological specimens in his *Arctic Zoology* (1784–85), but referred to their proprietor as the unnamed 'worthy and philosophical sister' of Ashton Blackburne. Whereas Ashton had 'added to the skill and zeal of a sportsman, the most pertinent remarks on the specimens he collected', Anna was portrayed as a passive beneficiary who had merely received the specimens he 'sent over' from America.[60]

Anna Blackburne eventually became relatively well known to other scholars of natural history thanks to the extent of her collections, her patronage and her own expertise. She offers an example of how a woman with substantive financial resources and supportive social connections might emulate the (male) model of amateur scholarship. Blackburne negotiated the potentially injurious associations of female scholarship by avoiding publicity for most of her life. She was by no means unusual in this respect, and indeed the point that is perhaps more remarkable is that Blackburne *did* eventually become known to the wider intellectual community, and thus to history.

Recent work by a number of historians has begun to uncover the varied histories of other learned women of the eighteenth century who, in accordance with contemporary notions of female propriety, modestly concealed their intellectual contributions from the public eye.[61] Just as for

---

See: Murray, *The British Garden* (1799), p. vi and George, *Botany*, p. 7. See also: Goodman, 'Suzanne Necker's *mélanges*', pp. 210–223, esp. pp. 215–219.

[58] Admittedly, the Botanical Society of Lichfield lauded her as one of three 'learned and ingenious Ladies' in the Preface to their translation of Linnaeus' *Systema vegetabilium* (1783), along with 'Mrs Egerton, *Oulton-Park, Cheshire*' and 'Mrs Cummins, *Kensington*'. Their public recognition of female intellectual prowess was exceptional, however. 'Preface of the Translators', in Linné, *A System of Vegetables* (1783), p. xii.

[59] Both father and daughter were collectors, although John devoted his attentions to his garden. For contemporary accounts of Orford, see: Aikin, *Description of the Country* (1795), pp. 307–308 and Pennant, *Tour in Scotland* (1776), p. 13.

[60] Pennant, *Arctic Zoology* (1784–85), vol. 1, p. A2; Wystrach, 'Anna Blackburne', pp. 156–157.

[61] See especially: Terrall, 'Frogs on the mantelpiece'; Cooper, 'Picturing nature', pp. 519–529; Dietz and Nutz, 'Collections curieuses', pp. 54, 64. Dietz and Nutz discuss Madame Dubois-Jourdain, whose Parisian collection may have rivalled that of the Duchess of Portland in size and scope. Dubois-Jourdain's cabinet is mentioned in Remy, *Catalogue*, pp. vii–viii, which was the sale catalogue for the Dézallier d'Argenville collections. Pierre Remy also described and sold the collection of another female collector, 'Madame de B***', in 1763.

commerce (discussed in Chapter 1), women's participation in Enlightenment science might be best considered within a 'familial model'. As it was socially unacceptable for most women to portray their learning in public, those who wished to participate in science tended to do so behind closed doors. Their contributions might, however, be represented externally by a male relative or head of household.[62]

The few women who openly emulated amateur scholarship would often attempt to counterbalance its 'masculine' associations by strongly emphasising their femininity in other arenas. Strategies included choosing to collect and study 'effeminate' objects and/or adorning their property or person in excessively 'feminine' styles. The Duchess of Portland collected shells, for example, which she arranged haphazardly rather than deploying the 'manly' taxonomic order of eighteenth-century scholarship.[63] Bluestocking Elizabeth Montagu decorated her London apartments in the flamboyantly feminine 'Chinese' taste, in an attempt to head off the perception that she had been 'unsexed' by her intellectual endeavours.[64] The few women who published books on science used 'feminine' genres such as poetry or the epistolary form, rather than deploying dry, 'masculine' description.[65]

The gendering of the connoisseur or amateur as male did not wholly exclude women from amateur science, but it did mean that female participation was never straightforward. Those women who emulated amateur scholarship did not usually question or resist the gendered norms that defined practice, behaviour and the judgement of value; instead, they would circumvent or negate the masculine associations that came with scholarship. The emergence of female amateurs was a major aspect of expanding public participation in science overall. Increasing participation led to the emergence of further apprehensions about amateur science, however. Men as well as women were subjected to censure over the ways in which they engaged with scholarship.

Anxieties about Amateur Science

Amateur scholarship was understood to foster a set of ideal behavioural standards among men. Even if the amateur was usually well thought

---

[62] Cooper, 'Homes and households'.

[63] Sloboda, 'Displaying materials', pp. 455–472.

[64] Sloboda, 'Fashioning Bluestocking conversation', pp. 129–148. One reason why Chinoiserie was associated with femininity was because displays of Chinese objects were perceived to be indiscriminate rather than systematic.

[65] George, *Botany*; Shteir, 'Priscilla Wakefield' and 'Botanical dialogues'.

of as a cultural figure, however, opposing associations also circulated – and these were formulated differently on either side of the Channel. In Britain, in particular, critics articulated sentiments that will seem very familiar to twenty-first-century readers: amateurs were portrayed as second-rate producers of substandard scholarship. The belittlement of amateur scholars on the grounds of superficiality should not be taken at face value, however. In both France and Britain, longstanding deliberations over who should partake in scholarly culture gained in urgency as participation in scientific learning widened.[66] Anxieties about public involvement in science were grounded in a cultural connection between scientific learning and social probity. Such concerns were often expressed through claims that amateur scholarship had diverged intellectually and culturally from 'mainstream' science.

Discussions of amateur scholarship in France were primarily focused on the fine arts rather than amateur study of the sciences. Contemporary writers paid attention to amateurs of art because they acted as intermediaries between the expanding market for artworks and scholars at the Académie Royale de Peintre et de Sculpture.[67] The power that amateurs of art accrued through this role had provoked heated debate about their legitimacy as scholars. In 1748, the Comte de Caylus had celebrated the amateur's refined judgement and good taste;[68] twenty years later, however, Denis Diderot lashed out at 'the damned race ... of amateurs',[69] claiming that they lacked proper knowledge about the subject and, being more interested in obtaining cheap artworks than in supporting true artistic genius, were stifling the quality of art.[70]

Amateurs were not condemned outright, however. Those engaged in the study of the natural sciences in particular were often subject to praise. The author of the *Encyclopédie* article 'Natural History' actually celebrated the amateur study of nature, explaining that, through their

[66] See references throughout this chapter, especially: Lynn, *Popular Science*, ch. 3; Sutton, *Science for a Polite Society*.

[67] Guichard, *Les amateurs*, pp. 110–112, 137–145, esp. p. 143. Guichard has revised Pomian's description of the relationship between traders and amateurs. She argues that amateurs of art were active as intermediaries between the market for artworks and the creation of collections. These amateurs supplemented the role played by tradesmen, because they also determined attributions and negotiated the prices of artworks on behalf of other collectors. Amateurs of botany had a more passive relationship with the marketplace than amateurs of art. They did not take on the same responsibility for negotiating prices for other collectors. For Pomian's argument, see: Pomian, *Collectionneurs*, pp. 168–184. Unlike in the world of art, the relationship between the market for plants and the collection of botanical specimens was negotiated by intermediaries such as botanically educated traders and gardeners.

[68] Guichard, *Les amateurs*, pp. 26–30.

[69] Diderot, 'Salon de 1767'; Guichard, *Les amateurs*, pp. 9–10.

[70] Diderot, 'Salon de 1767', p. 520.

collections, amateur naturalists 'contribute perhaps just as much' to natural history as more highly trained scholars.[71] Selectivity and specialisation were key to gaining a positive scholarly response: an amateur's well-stocked natural history cabinet was deemed an essential resource for all scholars, being 'an abridgement of nature in its entirety'.[72] By contrast, scholars increasingly condemned curiosity cabinets (which by the latter half of the eighteenth century were associated with indiscriminate collecting) as sites of spectacle and ostentation.[73]

The welcome that was extended to amateur natural history foundered, however, over the rocky problem of maintaining taxonomic clarity. André Thouin was typical of many when he wrote to Joseph Banks in 1784 that,

[I]t is dreadful to see the state into which Botany will fall, as heaps of Petty Authors who have only seen plants in books and who have never left their corner of the earth will make Systems, will change names [and] finally will overturn everything.[74]

Thouin feared that the profusion of practitioners might undermine botany because, by inventing names or even whole new classifications for plants, amateurs would confuse the work done by recognised taxonomic botanists. Like Diderot, he was not opposed to all forms of public participation in botany, and at other times he positively encouraged it.[75] But he stressed the need for a clear botanical authority which might impose standards for observation and naming among amateur botanists, thus increasing accuracy overall.

The challenge of policing intellectual quality among the mass of practitioners gained further urgency because of the moral and social significance attributed to scholarly culture. Other commentators expressed intense concern about the potential social and cultural consequences of amateur scholarship that they perceived as deficient. Representations of scholars who deviated from contemporary notions of acceptable

---

[71] Anon. [Daubenton?], 'Histoire naturelle' (1765), p. 228. Stemerding claims that the author was Daubenton (*Plants, Animals and Formulae*, p. 23), but the article was published anonymously and I have not found further evidence to support this.

[72] Anon. [Daubenton?], 'Histoire naturelle' (1765), p. 226; Pomian, *Collectionneurs*, pp. 132–138.

[73] Belhoste, *Paris Savant*, pp. 121–124, esp. p. 124.

[74] BL Add. Ms. 8095, f. 321, André Thouin (Paris) to Joseph Banks (London), 19 February 1784.

[75] See, for example: MNHN, Ms. 1934/XXX, 'Mémoire sur le Jardin du Roi. Rédigé par André Thouin en Octobre 1788', in which Thouin explains that, of the Jardin du Roi's four principal functions, its primary one was as a public garden for the working people of Paris.

Figure 3.1 Isaac Cruikshank, 'The Naturalist's Visit to the Florist' (London: Laurie & Whittle, 1798). Courtesy of the Lewis Walpole Library, Yale University.

behaviour appeared in literature,[76] the fine arts and, especially in Britain, in the burgeoning print culture.[77] In displaying aberrant forms of scholarship, these cultural manifestations reinforced the existing misconception that substandard scholarship was becoming a widespread problem.

One characteristic example is Isaac Cruikshank's 1798 print, 'The Naturalist's Visit to the Florist' (Figure 3.1), which depicts the calamitous effects of excessive enthusiasm for natural history collecting. A pleasant promenade to view the tulip collection of the overweight and

---

[76] For a selection from numerous literary examples, start with: George Colman's satirical journal *The Connoisseur* (published weekly in 1754); Coypel, *La Curiosimanie* (the play is undated but was written and performed between 1717 and 1747); Smollet, *The Expedition of Humphrey Clinker* (1771); Swift, *Gulliver's Travels*.

[77] Coltman, *Classical Sculpture*, ch. 5; Guichard, *Les amateurs*, pp. 126, 276–281, 301–305; Kelly, *Society of Dilettanti*, pp. 24–34; cf. Crow, *Painters and Public Life*, pp. 7–9.

gouty 'Florist' is interrupted when the 'Naturalist', who is equally large and ungainly, catches sight of a rare butterfly.[78] Immediately forgetting all decorum, the Naturalist whips off his wig (which he hopes to use to catch the tiny creature) and charges into the flowerbed, crushing the tender tulips beneath his feet. His friend, confined to a Bath Chair by gout, cries out in impotent horror. The butterfly, meanwhile, flies off into the clear sky above. The pair are counterbalanced visually by the passive figure of a bemused servant who pushes the chair, and by a gentleman who calmly watches from a distant seat in the wall. The latter's composure underscores the collectors' impropriety.

Cartoons lampooning natural history collectors, especially insect-fanciers and flower-collectors, became relatively common in Britain towards the end of the eighteenth century. Caricaturists articulated a moralising concern that the study of nature encouraged trifling, frivolous pursuits, far removed from social respectability. Anxieties about the social corruption that such kinds of behaviour could lead to were expressed, furthermore, in classed and gendered terms. Carried away by their fanatical zeal for collecting nature, the Naturalist and the Florist in Figure 3.1 have become aberrations of both culture and nature. Their ungentlemanly over-enthusiasm is condemned as a generator of behaviour alien to decent society. The fact that they are portrayed beside (and within) a bed of tulips is also significant – the collection of so-called florists' flowers was considered lower-class and effeminate, and therefore opposed to respectable, gentlemanly, polite scholarship. We will return to this point in the final section of the chapter. The pair appear almost monstrous, with their bald heads, bulging bodies and gaping mouths, which expose plebeian, ill-kept teeth.[79] Impotently chasing after fripperies, the collectors are no longer congruent with the natural and social worlds around them, demeaning both their social rank and their sex.

[78] The print is now in the National Portrait Gallery, NPG D12540. The caption suggests that the butterfly is an 'Emperor of Morocco'. This butterfly was frequently used in satires of substandard naturalists in the late 1780s and 1790s. Joseph Banks was a particular target, especially following his appointment as President of the Royal Society. For examples, see: Pindar, *Sir Joseph Banks* (1788) and Gillray, *Great South Sea Caterpillar* (1795). The fictional Emperor of Morocco was apparently based on the British native butterfly *Papilio iris* (L.). See: Bodleian, Ms. Douce c. 11, f. 68, Richard Twiss (Edmonton) to Francis Douce, 2 September 1794. It is unclear whether there was any connection between the satirical Emperor of Morocco butterfly and the gooseberries discussed in Chapter 2, which were given the same name. For more on the attacks on Joseph Banks, see: Bewell, 'On the banks of the South Sea', pp. 173–193.

[79] Open, toothless mouths were considered disgusting in the early modern period. Someone portrayed with such a mouth in the later eighteenth century was either plebeian or mad, or both. See: Jones, *Smile Revolution*, pp. 10, 54–59.

The condemnation of ungentlemanly amateur scholarship was posited on a cultural association between covetous collecting and an inability to control one's passions. The Naturalist and Florist in Figure 3.1 have been emasculated by their enthusiasm; the identification of their substandard scholarship with effeminacy expressed a longstanding connection between femininity and superficiality, lack of independence and lack of learning.[80] Especially in the latter half of the eighteenth century, however, a wealth of new images portrayed the opposite, showing male connoisseurs as 'oversexed', drooling over paintings or sculptures of naked female forms. While upper-ranking dilettanti and connoisseurs used libertine culture to assert their elite social status, the mainstream polite culture of the middling ranks was strongly opposed to such behaviour.[81] Nevertheless, cultural representations of connoisseurship fostered an association between collecting and excessive, unbridled lust.

Satirical portrayals of botanical culture throughout the eighteenth century regularly evoked an association between scholarship, flowers and sexuality. Gardens, and especially shrubberies, had long been literary locations for improper sexual behaviour;[82] images of women florists as sexual objects were simple extensions of this association. Figure 3.2, 'The Female Florists' (1773), depicts a woman and a girl dressed exquisitely in silk and pearls, with powdered hair and scarlet lips. The girl has been admiring a rose bush when the woman leans forward, takes her chin, and draws her face towards her. The subtext of the image is clear: the real object of admiration is the girl's beauty. The older 'florist', a courtesan, ominously assesses her innocent protégé; the rose bush that the girl tends symbolises her own virginity. A similar point is made in Figure 3.3, 'The Fair Florist' (c. 1780), in which a well-dressed lady ties a daisy-like flower to a stake. Once again, the object of the image is the 'florist' rather the flower. The text beneath reads: 'This fair ones [sic] glowing cheeks and sparkling eyes, Make her more lovely than the flowers she ties.' Both images draw on the longstanding association between flowers and female sexuality, and both assume a knowing male gaze. Connoisseurship functions here in two ways: in each, a female 'florist' works knowledgeably with her subject (the child in Figure 3.2, the pot-plant in Figure 3.3), seeking to enhance its beauty. But the real connoisseur is the (male) viewer, who will observe the activities depicted with

[80] Bending, 'Every man', p. 523; Bermingham, 'Elegant females', p. 492; Bhattacharya, 'Family jewels', pp. 207–226; Landes, *Women and the Public Sphere*, pp. 70–71; Larrère, 'Women, republicanism', pp. 140–141, 144–145.

[81] Kelly, *Society of Dilettanti*, pp. 30–35.

[82] Harvey, 'Gender, space and modernity', pp. 159–180.

Figure 3.2 'The Female Florists' (London: Robert Sayer, 10 January 1773).
Courtesy of the Lewis Walpole Library, Yale University.

Figure 3.3 John Raphael Smith, 'The Fair Florist' (print, London). Image ID: 00960914001 © The Trustees of the British Museum.

a knowing eye, proprietorially reading the messages about female sexual availability.

These three caricatures sum up some of the contemporary anxieties surrounding amateur botanical scholarship. We saw earlier that writers on polite scholarship such as Shaftesbury emphasised that their readers should guard against the excesses of specialisation and enthusiasm, and that (among the middling ranks at least) an education in science might reinforce standards for moral probity. Contemporary print culture lampooned the misguided vision, immoderate behaviour and poor taste of the supposedly ostentatious scholar;[83] such images created and reinforced wider anxieties about incompetent amateurs. Cruikshank's image of mad male natural history collectors suggests that the meaningless accumulation of insignificant natural objects would disrupt the social order by fostering both immoderate greed and social incapacity among practitioners. Figures 3.2 and 3.3 are more sinister, depicting the study of 'nature' as a source of, and cover for, moral degeneracy. Hyperbolic images such as these poked fun at a social figure that was easily recognisable, and gave the impression that defective amateur scholarship was much more widespread and problematic than it actually was.

Amateur Botany in Practice

Deficient amateur scholarship may have been less socially problematic than the caricatures and satires suggested, but eighteenth-century France and Britain nevertheless saw a range of amateur practices emerge. The polarised categories of expert 'lovers of botany' and ignorant 'careless collectors', and the caricatures of errant amateur scholarship, paper over a much greater diversity in expertise and interests among the wider public. Science education catered to (and so encouraged the development of) this assortment of approaches to botany. Exploring the range of practices that emerged offers an important corrective to characterisations of the eighteenth century as an age concerned purely with systematics and classification. For many amateurs, male and female, learning taxonomic botany was one element of a more heterodox approach to the science. Furthermore, it was viewed by many as kind of accomplishment that would give added distinction within social circles.

Paris and London sported a number of private gardens whose owners explicitly sought to make a commercial profit from the rising public

---

[83] Kelly, *Society of Dilettanti*, pp. 24–34.

interest in studying botany.[84] They offered educational support for botanical learning either in lecture courses or by encouraging amateurs to visit their gardens informally and converse with the proprietors. Numerous advertisements to this effect appeared in Paris in the 1770s and 1780s, from which we can construct a sense of the geographical distribution of these gardens, as shown on Map 1.2. The *Dictionnaire historique* (1789) explained that, 'in Paris there are a certain number of skilled Doctors, in addition to the Royal Professors, who devote themselves to teaching Botany, & who give public Courses at their homes, in their gardens'.[85] Public botany courses were, naturally, taught at the Jardin du Roi (location A on Map 1.2), but individual amateurs of botany also advertised courses from their own gardens. Tutors who publicised their courses in journals such as the *Avant-Coureur* and the *Journal de Paris* included the physician Barbeu Dubourg and a M. Royer, a *marchand-épicier-droguiste*, who gave annual botany courses in his gardens on the rue du Faubourg Saint Martin throughout the 1760s and early 1770s (G on Map 1.2).[86] Royer also allowed other botanists, including the royal surgeon Bergeret, to teach from his garden.[87]

The same was equally true across the Channel, where a range of entrepreneurs capitalised on the rising public interest in learning botany. One of the most telling examples is that of the London Botanic Garden (B and C on Map 1.1), founded by the business-minded apothecary and botanist William Curtis (1746–1799) on 1 January 1779. The garden, which was funded by public subscription,[88] was open for use by a range of publics that included, Curtis claimed, 'the Physician, the Apothecary, the student in Physic, the scientific Farmer, the Botanist, (particularly the English Botanist,) the lover of Flowers, and the Public in general'.[89] Curtis explained that he had received donations of

---

[84] Laissus, 'Les cabinets d'histoire naturelle', p. 665. For an overview of public natural history education in provincial France, see: Roche, 'Natural history'.

[85] Hurtaut and Magny, *Dictionnaire historique* (1789), vol. 1, p. 644.

[86] *Avant-Coureur Index*, vols 1–3 (1760–73), 1768, p. 295; 1773, p. 261. I am grateful to Dr E. C. Spary for allowing me to consult her notes from this Index.

[87] *Journal de Paris*, no. 180, 29 June 1785. I am grateful to Professor Bruno Belhoste for sending me this reference.

[88] In 1777, Curtis requested a 'loan of £50 for a year from a few of his Friends' to support the project, in addition to the subscription. NHM, DTC, vol. 1, f. 146, Sir Thomas Frankland (Stockeld, Nr Wetherby) to Joseph Banks, 28 August 1777; NHM, DTC, vol. 1, f. 154, Sir Thomas Frankland (Stockeld) to Joseph Banks, 14 November 1777.

[89] Curtis, *Proposals* (1778), title page. For two guineas a year, subscribers could use the gardens and their library between Tuesdays and Fridays, from 6am until 8pm. Curtis, *Proposals* (1778), p. 18.

plants for his garden from many places, especially 'His *Majesty's* match-less collection...at Kew'.[90] He also thanked commercial nurserymen 'Messrs Gordon, Lee, Kennedy and Malcolm', who had all given 'curi-ous exotics' to the garden.[91]

As we have seen, plant traders and amateur collectors presented their gardens as collections of interest to the wider public, and used jour-nals and guidebooks to publish announcements about their collections and encourage visitors to come and converse with them. As nursery-man Daniel Grimwood explained, 'a gentleman has liberty to walk in the nursery to look at such plants as he wants for his own use',[92] and the geographical concentration of the upper-end nurseries in London's northeastern and southwestern peripheries (indicated in Map 1.1) made it easy for amateurs to go 'Plant-hunting' in several at once. On one day in March 1791, for example, botanical amateur Richard Twiss (1747–1821) 'went with Sir John, Plant-hunting, to Hammersmith, Kensing-ton &c'. He described in detail the nurseries that he and his friend vis-ited, the rare plants they saw and what they purchased.[93] London's nurs-eries, like those in Paris, contained collections of new plants that rivalled those of private amateurs and public botanical institutions. Their exclu-sive contents made them significant as sites for new knowledge.

In addition to attending lecture courses and physically visiting gar-dens, amateur botanists also learnt through reading books on botany. The quantity of books produced for the wider public increased rapidly especially after 1760; by the nineteenth century, botany had became the most widely sold genre of natural history book in Britain.[94] The greater readability of botany books reflected a shift in science writing more generally. According to Maria Edgeworth, books of science pub-lished before mid-century 'were full of unintelligible jargon[,] and mys-tery veiled pompous ignorance from public contempt'. By 1798, how-ever, 'writers must [now] offer discoveries to the public in distinct terms, which every body may understand'.[95] The increased publication of well-written vernacular books had a dramatic effect upon public access

[90] Ibid., p. 10.   [91] Ibid., p. 10.

[92] OBP, 16 September 1795, trial of Charles Fairfield (t17950916–73). Grimwood worked at the nursery marked as #39 on Map 1.1.

[93] Bodleian, Ms. Douce d. 39, ff. 35–36, Richard Twiss (Bush Hill, Edmonton) to Francis Douce (Holborn), 26 March 1791. Twiss was a travel-writer and archetypical upper-middling British connoisseur. His interests ranged from chess to ancient history and from botany to papermaking. He was made a Fellow of the Royal Society in 1774 but was forced to withdraw in 1794 because of financial difficulties caused by a failed speculation on making paper from straw. Turner, 'Twiss, Richard'.

[94] Shteir, 'Botanical dialogues', pp. 301–303.

[95] Edgeworth, *Letters for Literary Ladies*, p. 20.

to science education, and particularly to botany.[96] The new botany books were of a high quality and mostly replaced Latin works, at least for the general public: even readers who 'had some Latin' apparently preferred botany books in the vernacular.[97] They were, moreover, affordable to those of a middling income. James Lee's *Introduction to Botany* (1760), William Withering's *Botanical Arrangement* (1776) and Jean-Jacques Rousseau's *Lettres élémentaires sur la botanique* (1781/1789) were among the most successful.[98]

Many of the botany books published after 1760 were directly addressed to female readers.[99] Women's participation in botany was complicated not only by the existing association between scholarship and masculinity, already discussed, but also by the morally dubious nature of Linnaean taxonomy. Linnaeus' system asked its pupils to classify plants by counting their sexual organs and then to describe the relationships between plants anthropomorphically as 'marriages'. Monogamy is, unfortunately, rather rare among vegetables, and so the Linnaean System appeared to be promoting a wide gamut of illicit sexual liaisons to impressionable female minds. Linnaean botany attracted a great deal of censure from moralising social commentators.[100]

The Linnaean System nevertheless gained a wide following in both Britain and France. It was strikingly easy to learn and apply, and most vernacular translations carefully neutralised the indecorous terminology. Noting that 'Botany in an English dress' had by 1776 'become a favourite amusement with the Ladies', William Withering removed Linnaeus' sexual language from his *Botanical Arrangement*, rendering

---

[96] Shteir, *Cultivating Women*, pp. 18–21.

[97] Horsman, 'Botanising in Linnaean Britain', pp. 3–4.

[98] Henrey, *Botanical and Horticultural Literature*, vol. 2, pp. 652–654; Shteir, *Cultivating Women*, pp. 18, 21–25. Note that Rousseau's *Lettres élémentaires*, composed between 1771 and 1774, circulated as manuscripts in Paris in the 1770s. These were published posthumously in Geneva in 1781. In 1785, the Professor of Botany at Cambridge, Thomas Martyn, published an English 'translation', which in fact considerably altered and supplemented Rousseau's original text, placing greater emphasis on the Linnaean System. Martyn's edition was then translated into French and published in France in 1789, and is responsible for a lasting, false impression that Rousseau himself advocated Linnaean botany. Cook, *Jean-Jacques Rousseau*, pp. 297, 308–318. In this chapter, I have used Martyn's English translation except where there are discrepancies between this and the edition published in 1781.

[99] Shteir, *Cultivating Women*, pp. 19–21.

[100] Women's participation in Linnaean botany was criticised in particular during times of wider social instability, for example in Britain during the Revolutionary and Napoleonic Wars. For more on the botanical controversies in the 1790s, start with: Bewell, 'On the banks of the South Sea'; Browne, 'Botany for gentlemen', pp. 593–621; George, *Botany*, pp. 105–152; Polwhele, *Unsex'd Females* (1798); Schiebinger, 'Private life of plants'; Teute, 'Loves of the plants'.

Amateur Botany in Practice

the taxonomic system suitable for the chaste eyes and minds of his readers.[101] The 1789 French translation of Thomas Martyn's Linnaean version of Jean-Jacques Rousseau's *Lettres élémentaires* likewise deployed neutral terminology.[102]

Reformist educational writers proposed that botany offered 'an antidote to the world of accomplishments' – a world which, according to Mary Wollstonecraft, encouraged irrationality and irresponsibility among women.[103] Girls were conventionally taught skills such as music, art and dancing, and learnt to perfect their dress and demeanour. The effect of this education, Wollstonecraft observed, was that 'strength of body and mind are sacrificed... to the desire of establishing themselves, – the only way women can rise in the world, – by marriage.'[104] Wollstonecraft, Hannah More and others deemed such an education superficial and pointless: these so-called achievements would not prepare women for their more important future duties as wives and educators of children.[105]

Presented as a 'rational' science, botany certainly offered a substantive contrast. But although it may have seemed a corrective to the worst failings of eighteenth-century female education, most women still applied their knowledge in ways that conformed to existing cultural norms and reinforced conventional ideas about femininity. After all, the cultivation of accomplishments remained an important part of a polite female identity until well into the nineteenth century, despite censure from reformist writers such as Wollstonecraft and More. Botanical learning may have provided a remedy for the superficiality that was associated with many accomplishments, but it did not replace accomplishments themselves. Most women applied their knowledge by deploying the skills they had learnt earlier in life.

---

[101] Withering, *Botanical Arrangement* (1776), vol. 1, p. v.

[102] Rousseau [and Martyn], *Lettres élémentaires* (1789). As Ann B. Shteir and Alexandra Cook have shown, Thomas Martyn's English translation of Rousseau's *Lettres* gave them 'a Linnaean slant... that differed from Rousseau's own', which was only introduced to France following the French translation of Martyn's English edition in 1789. Rousseau's original *Lettres* deploy Jussieu's natural family system in addition to that of Linnaeus. See: Shteir, *Cultivating Women*, quote p. 19. See also: Cook, 'Le pluralisme taxonomique', pp. 17–22, *Jean-Jacques Rousseau*, pp. 167–172, 308–318 and 'Languages of music and botany', pp. 85–86.

[103] The quotation is from Shteir, 'Botanical dialogues', p. 304. For Wollstonecraft's discussion, see: *Vindication of the Rights of Woman* (1792), pp. 112–113.

[104] Wollstonecraft, *Vindication of the Rights of Woman* (1792), p. 112.

[105] Ibid., pp. 138–139, 164–165, 279. See also: Bermingham, 'Elegant females', pp. 491, 497; Campbell Orr, 'Aristocratic feminism', pp. 317–320; Vickery, 'Theory and practice', pp. 94–109.

Some of the better-known ways in which scientific knowledge and women's accomplishments were linked was through producing representations of flowers in botanical painting, through embroidery or tapestry and through gardening or garden design. Within this range, the 'paper mosaicks' created by Mary Delany (1700–1788) constitute particularly striking examples of how botanical knowledge could be integrated within the female culture of accomplishments. Figure 3.4 depicts a bay-leaved passionflower, which Delany created circa 1776. The flower is formed through a collage of tiny strips of paper, each individually cut, coloured and glued on to the background.[106] Each collage features the exact number of stamens and styles in the flower, with its Linnaean classification noted on the back.[107] The botanically accurate collages were celebrated by contemporary botanists: Joseph Banks declared (somewhat hyperbolically) that they were the 'only' representations of nature he had ever seen from which he could describe any plant botanically.[108] Delany received specimens to copy from Kew and Chelsea, from the gardens of amateurs such as John Fothergill and William Pitcairn, and from Kennedy and Lee's Vineyard Nursery.[109]

Delany's representations of plants do not now conform to current botanical conventions, and are today often dismissed as 'quaint' rather than as 'serious' applications of learning.[110] Her work is, however, a wonderful example of how botanical knowledge could be valued for reasons that had more to do with gendered notions of acceptable social and scholarly behaviour than strict scientific conventions. Mary Delany's collages show how botanical knowledge might be applied in a way that accorded with 'female' accomplishments as well as with 'male' taxonomic botany.

By using botanical knowledge to enhance and display their existing accomplishments, middling and upper-ranking women found means of respectably putting their botanical learning to practical use. The profusion of accessible botany books published in the latter half of the eighteenth century meant that a student of botany could develop a strong grounding in botanical taxonomy and nomenclature. For female amateurs of botany, such knowledge might then be deployed acceptably in creative or imaginative ways that were not necessarily aligned with the original taxonomic intentions of botanical authorities.

---

[106] The flower alone is composed of around 230 paper petals. Hayden, *Mrs Delany*, pp. 51, 67–68, 132–134. For more on Mary Delany, start with: Delany, *Autobiography* and the contributions to Laird and Weisberg-Roberts, *Mrs Delany and Her Circle*.

[107] Hayden, *Mrs Delany*, pp. 143–146.   [108] Ibid., p. 158.

[109] Delany made 'mosaicks' of eighty-four specimens sent from Kew and fourteen specimens from Chelsea. Ibid., pp. 135–136, 153–154.

[110] Blunt, *Art of Botanical Illustration*, p. 155.

Figure 3.4 Mary Delany, '*Passiflora laurifolia*: bay leaved' (paper collage). Image ID: 00037142001 © The Trustees of the British Museum.

Taste

The heterodox practices that emerged within the amateur study of botany were not specific to female botanists. Male amateurs could also rebuff the emphasis on taxonomy that was laid down by botanical authorities. This was particularly so if the amateur botanist (male or female) was interested in the science of gardening. As a number of botanical and horticultural texts published in the late eighteenth century suggest, the connection between amateur botany and gardening was underpinned by the long-standing ideas about scholarly connoisseurship previously sketched out. Particular emphasis was placed on cultivating good taste and on training the vision; the refinement of both attributes was considered by many amateurs to be as significant as naming and classifying the plants themselves.

The emphasis on taste within the sciences was most obviously made manifest in contemporary expectations about visual elegance. The rationale for the arrangement of eighteenth-century natural history books, and the display of objects in collections of *naturalia*, was strongly influenced by expectations about 'good taste'.[111] Likewise, collections of dried plants in herbaria had to be 'tasteful' and 'pleasing', as explained by Jean-Jacques Rousseau in his *Lettres élémentaires*. Care was needed, Rousseau emphasised, not to 'spoil and disfigure a collection'.[112] The possession of good taste was also necessary for garden and landscape design and was equated, too, with the proper management of land. Writing about the management of private (non-commercial) nursery gardens in the *Jardinier Portatif*, Thomas-François de Grace explained that '[t]he true secret of having good nurseries consists of keeping them well maintained:... that demands a little care and [imposes] some constraints; but... one should see this work as a pleasure that we taste [*qu'on goûte*] in governing them.'[113]

But what was 'taste'? The multi-authored article 'Taste' in the *Encyclopédie* anatomised its component parts:

In order to have taste, it is not enough...to see [and] to know the beauty of a work; one ought to feel it, and be moved by it... [O]ne must discern the different shades [of feeling]; nothing should escape an immediate perception... the man of taste, the connoisseur, will see in one rapid glance the mixing of styles.[114]

---

[111] Dietz and Nutz, 'Collections curieuses', pp. 57–60, 66; Spary, 'Scientific symmetries', pp. 1–46; van de Roemer, 'Neat nature', pp. 47–84.
[112] Rousseau, *Letters on the Elements of Botany* (1787), p. 79.
[113] Anon., *Jardinier Portatif*, p. 19.   [114] Voltaire *et al.*, 'Goût' (1757).

Taste was an attribute that was internalised; a person (usually a man) in possession of good taste was a connoisseur. The *Encyclopédie* article continued by drawing a distinction between gastronomy, which it described as 'sensual taste', and 'intellectual taste', which 'needs more time to develop':

Man moulds and educates his taste in art much more than his sensual taste... A young man who is sensitive but untutored cannot at first distinguish the parts in a large chorus; in a painting, his eyes do not at first distinguish the shadings, the chiaroscuro, the perspective, the harmony of its colours, and the correctness of the draughtsmanship; yet little by little his ear learns to hear and his eyes to see.[115]

Enlightenment theories on taste explained that it was the product of a sensitive nature and a careful education. Men (and a small number of women) who aspired to educate their taste and become a connoisseur focused primarily on learning to appreciate the fine arts. They were expected to gain an intellectual understanding of aesthetic theory and to refine the senses: connoisseurship was based on a sensory epistemology; a tasteful education involved training the eye and ear through practice and observation, and through reflecting intelligently on what had been seen or heard.[116] An amateur learnt to prioritise visual or aural evidence, through which his or her aesthetic judgement would be formed.[117]

The attribute of good taste was also understood to contribute to the moral development of the student. Drawing, for example, was 'a disciplining of the hand by the mind.'[118] As Joshua Kirby (tutor in perspective drawing to George III) explained, his students would be 'taught to SEE' with 'exactness' and 'judgement'.[119] A correctly trained vision would lead to a disciplined control over the passions, and a refined accuracy of judgement.[120] Students of botany were likewise taught to view nature in a particular way. Rousseau's *Lettres élementaires* underlined the significance to amateur botany of a correctly trained vision: 'Before we teach [novice botanists]...to name what they see, let us begin by teaching

---

[115] Ibid.
[116] See also: Bleichmar, 'Learning to look', pp. 85–111; Spary, 'The "nature" of Enlightenment', pp. 293–296.
[117] Dietz and Nutz, 'Collections curieuses', pp. 50–54; Pomian, *Collectionneurs*, pp. 170–172. Guichard emphasises the significance to amateurs of 'training the eye'. Guichard, *Les amateurs*, pp. 173, 178–185.
[118] Bermingham, *Learning to Draw*, p. 45.
[119] Joshua Kirby, *Dr Brook Taylor's Method of Perspective Made Easy* (1765), p. vi, quoted in Grant, 'Mechanical experiments', p. 204.
[120] Grant, 'Mechanical experiments', p. 210.

them how to see.'[121] Rousseau explained to his readers that 'it is not the nomenclature of a parrot which I wish you to acquire, but a real science, and one of the most delightful sciences that it is possible to cultivate'.[122] Rather than presenting botany as 'a mere labour of the memory', then, Rousseau framed the science as 'a study of observations and facts truly worthy of a naturalist.'[123] His aim was that his students would develop a deep understanding of nature through their observations. An apprentice botanist had to learn 'how to find out things of herself', and this was only possible if she learnt to view nature 'correctly'.[124]

The would-be connoisseur of the fine arts was ideally taught to see and understand the close geometrical detail necessary for perspective drawing, as well as how to appreciate the whole from a larger perspective.[125] As Buffon had explained in his *Premier Discours*, a youthful amateur of botany (or of any natural science) should first see and inspect his or her subject closely and then expand outwards gradually, so that he or she would eventually appreciate the connections between multiple parts of the natural world.[126] The principles underpinning connoisseurship of the fine arts and the science of botany are not identical, of course (and the two remained distinct from each other). But what the evidence does suggest is that the emphasis placed in the fine arts on cultivating a polite student's fine sensibilities and on developing their taste carried over into educational writings on botany. Botanical writers all emphasised the significance of developing the understanding rather than simply absorbing new information, and they often linked this directly to the cultivation of the student's aesthetic sensibilities.

Writers on botany and horticulture suggested that only a correctly trained amateur eye would be able to see and appreciate the individual qualities of each plant. The French texts articulated this connection very clearly. Successive editions of the *Bon Jardinier* proposed that plants should be examined carefully by the 'curious', and it recommended using a *loupe à la main* (magnifying glass) or *lorgnettes* (spectacles). A plant such as the geum (*Geum*, L.), for example, was 'very pretty', but

---

[121] Rousseau, *Letters on the Elements of Botany* (1785), p. 48. Rousseau's emphasis on the visual aspect of botany was replicated in other books of botanical instruction, notably Wakefield's *Introduction to Botany* (1786). See: Shteir, 'Priscilla Wakefield', p. 31. On training the eye of the amateur natural historian more generally, see: Terrall, *Catching Nature*, ch. 4.

[122] Rousseau, *Letters on the Elements of Botany* (1785), p. 60. Note that Martyn deviated from the original French in mentioning parrots, but the text is otherwise the same. C.f. Rousseau, *Lettres élémentaires* (1782), p. 557.

[123] Rousseau, *Letters on the Elements of Botany* (1785), p. 25.   [124] Ibid., p. 26.

[125] Grant, 'Mechanical experiments'.   [126] Buffon, 'Initial discourse', p. 99.

one needed 'a magnifying-glass to see all its beauties'. Similarly, the frax-
inella (*Dictamnus albus*, L.) 'has absolutely no effect in a Parterre; but
seen up close, has something that would flatter a *Curieux*'.[127]

But did the close visual examination recommended by the *Bon Jardinier* actually count as botany? The extensive provision of botanical education offered in the botany books and lecture courses mentioned before
shows that amateurs enthusiastically learnt taxonomic botany. They had
no particular need to understand botanical classifications, however, if
they simply sought to undertake the examinations proposed by the *Bon
Jardinier*. The two approaches to 'botany' were connected by assumptions about the value of a visual education derived from the connoisseurship of art. Both approaches found a home in the application of
botanical knowledge, especially in the context of 'scientific' gardening, or
horticulture.

Amateurs and the Science of Gardening

Expectations about how much botany an amateur should know developed in relation to the emerging science of horticulture. Gardening was
celebrated in contemporary books on horticulture as a form of 'knowledge [*connaissance*]' and 'a gracious amusement';[128] it was a suitable pastime for polite society because it was a refined pursuit, which congenially
combined aesthetic expertise with a knowledge of plants, and which
directly connected to the cultivation of taste. A scientific knowledge of
plants is not, of course, actually necessary in order to cultivate a garden,
but the two were repeatedly presented as counterparts of each other,
especially in books on botany written for the wider public. In 1759,
the Chelsea gardener Philip Miller used the phrase-names from Linnaeus' *Species Plantarum* in his hugely successful *Gardener's Dictionary*,
for example, and in 1768 he fully converted the classification used in the
Dictionary from that of Tournefort to Linnaeus.[129] 'Botany', declared
William Withering in 1776, 'is not to be learnt in the closet; you must
go forth into the garden or the fields, and there become familiar with

---

[127] Anon. [De Grace?], *Bon Jardinier* (1768), pp. 112, 106. See also: Rousseau, *Lettres
élémentaires* (1782), pp. 537 ('œil clairvoyant'), 541 ('loupe'), 556 (difficulty of seeing
glands).
[128] De Grace, *Bon Jardinier* (1768), pp. xvii, xix.
[129] Sixteen editions of the *Gardener's Dictionary* were published between 1735 and 1775;
they were a conveniently small size, which meant that they could be taken with ease
into the garden or field. Priced at 4s. 6d., they were not expensive. Botanical and
horticultural information thus gradually came within reach of even the most modest of
plant collectors. Henrey, *Botanical and Horticultural Literature*, vol. 2, pp. 633, 652.

Nature her-self.'[130] French writers explicitly celebrated gardening along with the scientific study of plants, perceiving that doing one usually led to an interest in the other. François Lebreton, the author of a successful *Manuel de Botanique* (1787), explained that, in France, 'The taste for *jardins anglais* has led to that for the culture of trees, the desire to know these trees systematically has come next, and this desire has spread itself over all parts of the vegetable kingdom.'[131] Le Berryais' *Traité des jardins* of 1789 thus began with simplified explanations of botanical subjects related to gardening, and later provided a lexicon of 'botany and gardening terms'.[132] The close connection between the science of plants and the practical expertise and aesthetic knowledge required for horticulture resulted in considerable crossover between the contents of horticultural texts and of those that were ostensibly botanical.

The promotion of gardening was in part a response to concerns about superficial engagements with the science. It was also an expression of the contemporary emphasis on contributing to the public good. Amateurs of botany were repeatedly encouraged to garden because gardening was a source of useful practical botanical knowledge quite different to botanical theorising. The *Encyclopédie* article 'Natural History' (1765) noted that, '[t]he cultivation of ornamental flowers and trees belongs to Botany', but ornamental gardening was nevertheless perceived solely as an 'innocent amusement'.[133] It stressed, then, the benefits that might be derived from a more structured approach to the study of flowers:

> If we had considered thoroughly this successive order of natural tints in flowers, if we had properly observed the leaves of holly and of other trees with variegated leaves, we would be able to draw from them new ideas for the mixing of colours in the arts, for the changing of these colours, ... etc. Such knowledge would be all the more certain because it would be in accordance with the operations of nature.[134]

Gilbert White, author of the runaway bestseller *Natural History of Selbourne* (1789), similarly encouraged his readers to broaden their knowledge from botanical systems to encompass an understanding of plant growth and cultivation:

> The standing objection to botany has always been, that it is a pursuit that amuses the fancy and exercises the memory, without improving the mind or advancing

---

[130] Withering, *Botanical Arrangement* (1776), p. xi. See also: BL, Add. Ms. 8096, f. 550, Benjamin Waterhouse (Cambridge, America) to Sir Joseph Banks (London), 10 August 1787.

[131] Lebreton, *Manuel de Botanique* (1787), p. vi.

[132] Le Berryais, *Traité des jardins* (1789), vol. 1, p. 3; vol. 2, pp. 433–443.

[133] Anon. [Daubenton?], 'Histoire naturelle', p. 227.   [134] Ibid.

any real knowledge: and where the science is carried no farther than a mere systematic classification, this charge is but too true. But the botanist that is desirous of wiping off this aspersion should be by no means content with a list of names; he should study plants philosophically, should investigate the laws of vegetation, should examine the powers and virtues of efficacious herbs, should promote their cultivation; and graft the gardener, the planter, and the husbandman, on the phytologist.[135]

Gardening, then, was a central component of useful botanical amateurship.

The profusion of expensive, beautifully illustrated books that were seemingly about botany offers further evidence for the integration of amateur botany and a culture of flower appreciation. We might assume from their titles that books such as John Hill's *Exotic Botany* (1759) or Conrad Loddiges' *Botanical Cabinet* (1817) were works about the latest botanical innovations. But the books, which were explicitly written by and for 'amateurs', in fact paid only cursory attention to botanical taxonomies. Perhaps better titled 'floral' (rather than 'botanical') books, they were lavish picture books that portrayed some of the flashiest exotic ornamental specimens. They featured a wide range of plants, which were arranged according to their Linnaean taxonomic class. But the authors drew readers' attention to flowers that were aesthetically appealing. John Hill's *Exotic Botany*, for example, emphasised plants prone to floral mutations known as 'double flowers', in which additional petals grow in place of sexual organs.[136]

Exotic Botany also gave its readers a basic introduction to Linnaean taxonomy. Writing about the 'Profuse Nyctanthes'[137] (Figure 3.5), a plant recently introduced from China, Hill explained its position within Linnaeus' 'Diandrya' class but then noted that,

It will be worth while to examine this Flower strictly, for the sake of that which follows. I know no Subject more Curious than searching Nature in her Course of doubling Flowers: and this is at once a singular and very glorious Instance. In many others the Filaments swell into Petals, and the Doubleness begins from the Base of the Flower; in this the Luxuriance rises from the Head of the Tube, and the two small Filaments remain unalter'd at its Bottom. This Flower of nine Petals is an approach to Doubleness; and will lead toward the Knowledge of the other.[138]

---

[135] White used 'phytologist' to refer to the taxonomic botanist. White, *Natural History* (1789), p. 183.

[136] Hill, *Exotic Botany* (1759).

[137] This plant was probably the *Nyctanthes sambac*, as described by Linnaeus in his *Species Plantarum* of 1753. *N. sambac* was transferred to the *Jasminum* genera by William Aiton in the *Hortus Kewensis*, vol. I (1789), p. 8.

[138] Hill, *Exotic Botany* (1759), p. 2.

Figure 3.5 'Profuse Nyctanthes', from John Hill, *Exotic Botany*, p. 2. Courtesy of the Lewis Walpole Library, Yale University.

In other words, Hill established the plant's place within Linnaean taxonomy but then extended the discussion to explain how double flowers occurred. Elsewhere in the book, he noted how readers might encourage this phenomenon among their own specimen collections.

Hill's integration of Linnaean botany with the art of breeding florists' flowers seems surprising. Although 'abnormal' flowers are now subjects of scientific interest, particularly among geneticists,[139] eighteenth-century taxonomic botanists dismissed the floral mutations as 'monstrous' aberrations that fell outwith the parameters of botanical science. Taxonomic classification, after all, requires 'type' specimens – plants with stable features that represent their *genus* – and horticultural specimens were therefore banished from taxonomic botany. Equally, as the production of double flowers had no obvious application in agricultural improvement, such floral knowledge would appear unequivocally useless. Nevertheless, these floral botany books presented florists' flowers as an enduring aspect of botanical scholarship.

Historians have generally followed the writings of eighteenth-century botanical authorities and considered the history of flower breeding as something separate from the scientific study of plants and the application of botanical knowledge. But the readers of the floral books and the customers who purchased exotic plants from the upper-end nurseries persisted in considering the production of unusual blooms to be part of the culture of botany.

Useful Floriculture

A further reason why botanical authorities dismissed florists' knowledge was because it was ingloriously associated with the lower classes. Mary Wollstonecraft, for example, declared florists' flowers to be 'the playthings of those deficient in taste or breeding'.[140] Eighteenth-century flower breeding was (and is) often portrayed as an activity that was specific to the culture of artisans, who formed 'florists societies' and competed over growing flamboyant blooms such as tulips, auriculas, pinks and primroses.[141] However, the culture of floristry was not nearly as distinct from eighteenth-century practical botany as historians have assumed.

[139] Meyerowitz *et al.*, 'Abnormal flowers and pattern formation', pp. 209–217.
[140] Wollstonecraft, *Original Stories* (1788, 1791), quoted in George, *Botany*, p. 156.
[141] Duthrie, 'English florists' societies', pp. 17–35; George, *Botany*, pp. 156–165. See the work of Anne Secord for more on artisans' participation in botany in the late eighteenth and early nineteenth centuries, esp. Secord, 'Artisan botany', p. 383 and 'Science in the pub', pp. 269–315.

Eighteenth-century taxonomic botanists did not publicly draw on gardeners' knowledge. Philip Gilmartin has discussed observations made about 'heterostyly' in the genus *Primula*, in which individual plants of the same species produce either two long stamens and one short pistil or two short stamens and one long pistil. Although Charles Darwin was the first to comment in print on this variation in the nineteenth century, Gilmartin's study of eighteenth-century botanical plates shows conclusively that the variation was well known to gardeners in that century. Preoccupied with questions of taxonomy, however, eighteenth-century botanists had not considered the variation to be botanically significant.[142]

The construction of floristry as lower-class and unscientific by botanical authorities has had lasting cultural resonance, but the extent to which floristry was valued, and the ways in which it was linked to botany, actually varied greatly within social groups. We have seen that the nurseries and some botanical amateurs unproblematically integrated floristry into their botanical studies, presenting the pair as mutually supportive. They did so in spite of the association between lower-class vulgarity, florists' flowers and gardeners' knowledge.

Indeed, evidence that florists' flowers were in fact widely appreciated across the social levels comes (perversely) from the increasingly fierce denigration of floristry by some writers on botany. In France, for example, Jean-Jacques Rousseau dismissed cultivars as 'monsters' and, true to form, linked flower breeding to corruption within civil society more generally.[143] In Britain, the debate about florists' flowers became caught up in the wider discussion of women's participation in botany. The poet and novelist Charlotte Smith (1749–1806), for example, repeatedly contrasted simple-minded female florists to proficient women botanists in her literary works, as did Mary Wollstonecraft, Maria Jacson and Priscilla Wakefield.[144] These authors singled out floristry as deficient and degenerate, and as something quite distinct from respectable botany. Their attack on female floristry was intended as a defence of women's participation in botany. The strength of their critique of floristry suggests, again, that cultivating florists' flowers was in fact widely practised among the middling and upper ranks.

Floristry, moreover, was not universally condemned by advocates of botany. The culture of amateurship – which blended taste, connoisseurship and scholarly study – offered a means by which some commentators

---

[142] Gilmartin, 'On the origins of heterostyly', pp. 39–51.
[143] Rousseau, *Lettres élémentaires* (1782), p. 579.   [144] George, *Botany*, pp. 162–164.

made the collection of florists' flowers culturally acceptable. This was especially so among urban communities in France, where amateurship in any case enjoyed a more positive cultural resonance. One example is a series of letters to the agricultural journal, the *Feuille du Cultivateur*, printed in 1793 (the timing is rather surprising, given that France was then descending into Terror). Writing about the collection and cultivation of rare and precious flowers, the former Marquis de Gouffier[145] assumed that his amateur readers would collect both 'florists flowers' and rare exotics. 'Without doubt', he wrote, 'a striking tulip, a pretty pink, [or] a beautiful buttercup, are of equal merit to [...] a rhododendron [or] an azalea.'[146] Gouffier's incorporation of florists' flowers into his promotion of amateur botany suggests that the 'botany' promulgated by taxonomic botanists was not the same as that practised by the eighteenth-century urban amateur public.

These examples reveal a history of considerable engagement with 'lower', more practical forms of knowledge among the middling and upper ranks. The fact that such knowledge was granted inferior status by botanical authorities meant that such experimentation was not represented as 'botany' in the eighteenth century. Indeed, the author of the *Encyclopédie* article 'Natural History' admired the knowledge developed among flower breeders because of its specifically practical application. 'Florists', the author wrote, 'know how to distinguish, from among tulips of different colours, those whose seeds will produce variegated tulips, and they anticipate the changes in colour that will occur each year in these variegated flowers.'[147] As flower appreciators, amateurs of botany were socially and culturally distinct from the artisan florists and gardeners. Like them, however, they relied on and developed practical horticultural skill.

Amateurs were important, then, as a group that encouraged experimentation upon, and breeding of, ornamental plants. The *Encyclopédie* article 'Natural History' continued by expressing regret that further practical botanical information had *not* been developed from the study of flowers:

---

[145] The Marquis de Gouffier was a noted amateur with property on Rue d'Aguesseau and Faubourg Saint Honoré, and at Pierrefitte-sur-Seine. He exchanged ten letters with André Thouin between 1786 and 1788 (MNHN, Ms. 314, Etat de la correspondence d'André Thouin). The Gouffier family was one of France's ancient noble families, one branch of which was joined with that of the Choiseul dynasty.

[146] *Feuille du Cultivateur*, 30 Janvier 1793, pp. 38–40; 2 Février 1793, pp. 42–44, quote pp. 43–44.

[147] Anon. [Daubenton?], 'Histoire naturelle' (1765), p. 227.

The cultivation of flowers [as undertaken by florists] requires very assiduous care; one must be attentive to the nature of each plant in order to prevent the diseases to which it is subject, and to prevent the plant from deteriorating; thus one is able to recognize, so to speak, the different qualities of their temperament, their hereditary diseases, and other characteristics of the plant kingdom.[148]

The contemporary emphasis on marrying botanical and practical knowledge indicates the existence of an epistemological subculture among amateurs of botany. The focus of this subculture was neither on the development of taxonomy nor on the useful application of that knowledge for agricultural improvement. Instead, it was concerned with the development of an aesthetic appreciation of plants underpinned by taxonomic botanical understanding. Amateurs of botany were important as a group that encouraged experimentation upon, and breeding of, plants (knowledge which might later be applied in more utilitarian contexts). The amateur appreciation of flowers was, in effect, a middle- and upper-ranking reformulation of lower-class floristry, made socially acceptable by its gloss of 'science'.

Most writers on horticulture unthinkingly absorbed assumptions about social status, claiming that the art of flower breeding had remained underdeveloped until the upper ranks began to patronise it. Gilbert White's assertion to this effect was absolutely typical: '[i]t was not till gentlemen took up the study of horticulture themselves', he wrote in 1789, 'that the knowledge of gardening made such hasty advances'.[149] Assumptions about status distinctions hampered communication between social groups, as did the values placed on different kinds of knowledge.

The conceptual connection between botany and the study of floriculture went further than encouraging the collection and cultivation of such plants by the middling and upper ranks. Somewhat surprisingly from a modern perspective, contemporary writers suggested that an amateur's refined vision would endow him or her with superior *practical* ability as a cultivator. In 1793, citoyen Gouffier underlined the importance of the possession of good 'taste', not just for the appreciation of flowers, but also for their cultivation. Noting the number of exotic trees in and around Paris (thanks to the fashion for *jardins anglais*), he complained that their wealthy owners had abandoned them to the care of gardeners. The effect of this neglect was even worse for exotic flowers, which Gouffier believed would not survive unless cultivated under the attentive eye of an amateur. This, he explained, was another reason why Paris lacked collections of exotic plants:

[148] Ibid.   [149] White, *Natural History* (1789), p. 177.

[I]f one does not love them [flowers] sufficiently to cultivate them oneself, or if one cannot give them enough time, one will never come to form or to sustain a beautiful collection; because how [can one] expect from a gardener, who habitually works more by duty than by taste, the minute cares that elegant flowers [*fleurs distinguées*] require, which, for reason of their beauty, are the most delicate? They need at the very least the clear-seeing eye of the master.[150]

These tender plants required more care and attention than a normal gardener would be willing or able to give. By contrast, someone in the possession of good taste was uniquely equipped to cultivate rare specimens.

Gouffier suggested, then, that the amateur was distinguished over and above others who worked within the garden by his or her particular ability to see. A tasteful engagement with the natural world would allow the amateur to perceive and attend to the tiniest needs of each specimen. Other writers made a similar differentiation between the sensitive, educated and clear-seeing amateur and the other people who worked in gardens. Reflecting in 1803 on an article about landscape gardening in the *Critical Review*, one Northumberland gentleman distinguished between the capacities of judgement possessed by the 'man of Science & of Taste' who possessed a landscape garden, and those of the various workers who managed it or who supplied specimens.[151] Concerned with the arrangement of trees within his grounds, the gentleman noted that:

[T]he Effect will be very different whether the axe be committed to the hand of Genius or the paw of avarice[.] The land steward ... w^d mark the trees of full Growth & ... fit for immediate use or separate those which he thinks too near each other but the man of Science & of Taste will search with scrutinizing care for Groups & combinations such as his memory recalls in the pictures of the best masters ... he will discover Beauties in a Tree which the others w^d condemn for its Decay.[152]

Good taste was considered necessary for the arrangement of plants in gardens and for their cultivation. But taste was not only a question of observation; it was deemed also to have a practical application in the cultivation of specimens. Gouffier and Adams explicitly separated the ability of an amateur to care for rare plants from that of the gardener or steward.

---

[150] *Feuille du Cultivateur*, 30 Janvier 1793, p. 39.

[151] The writer was Thomas Adams of Eshott Hall, Northumberland. Adams was a solicitor and agent for the Duke of Northumberland. He had purchased Eshott Hall in 1792, and set about improving the estate and its gardens. Many of the Eshott Hall papers are held at the Lewis Walpole library, Yale University.

[152] Lewis Walpole Library, Eshott Papers Mss 2, Box 24, [Thomas Adams?] Note entitled '1803 in British Crit^l Rev^w. Repton on Landscape Gardening & architecture.'

A tasteful education equipped an amateur of botany with a capacity to see flowers' finest distinctions and to tend to their needs accordingly. Amateurship incorporated, but was about much more than, learning taxonomic botany. Framed by a scholarly culture that emphasised above all the formation of good taste, an amateur of botany learnt how to inspect plants closely, how to arrange them for visual effect in a garden and how best to cultivate them. Ideally, practical gardening knowledge would be combined with an understanding of botanical taxonomy or nomenclature. Amateurs of botany developed new forms of knowledge derived from practical participation in gardening. This, in the nineteenth century, became known as the science of horticulture.

In a letter dated 14 December 1796, James Lee (junior) announced to a correspondent that 'we have raised the *azalea pontica* from seed. A plant all amateurs have been longing for these fifty years ... [it] will be very dear & possibly will be for sale in Spring.'[153] Commercial plant traders engaged with, and followed, the wishes and wants of customers who identified as amateurs, seeking to drive up the potential profits that they could make from selling them rare exotic plants. Eighteenth-century amateur science encompassed a much greater range of behaviours and levels of expertise than Thornton's two categories of expert 'lovers of botany' and ignorant 'careless collectors' might suggest.

The cultural formulation of the 'amateur' suggested a means of engaging with the natural world that was framed by polite science and firmly situated within Enlightenment scholarly culture. The latter absorbed and promoted forms of 'gentlemanly' comportment which placed particular emphasis on politeness and on a longstanding culture of scholarly connoisseurship. According to Enlightenment writers, polite French and British scholars should ideally avoid specialisation; amateurs of botany would therefore study the science of plants among a range of other subjects.[154] They should 'love' their subjects of study but should restrain themselves from giving way to unbridled or excessive enthusiasm.[155] Most importantly, they should cultivate a highly developed aesthetic judgement by refining their sensibilities. This was summed up as the sense of 'taste' and was the connoisseur's most cherished attribute.

The injurious contemporary depictions of eighteenth-century amateur botany (and amateur science in general) have misleadingly presented the non-professional study of plants as trivial and inconsequential. Portrayals

---

[153] BL, Add. 29533, f. 59, Letter from James Lee (Hammersmith), 14 December 1796.
[154] Brockliss, *Calvet's Web*, esp. pp. 14–17; Gascoigne, *Joseph Banks*, pp. 58–60, 66.
[155] Benedict, *Curiosity*, pp. 5–9, 18–19; Daston and Park, *Wonders*, pp. 306–328; Shapin, 'A scholar and a gentleman', pp. 300–303.

of deficient scholarship commonly presented the amateur as excessively specialised or disproportionately enthusiastic, or as engaged in depraved activities. Such depictions drew on wider concerns about the negative effects that too much specialisation or enthusiasm, or a lack of understanding, could have on an individual or on science at large.

Amateur botany encouraged an interest in practical gardening and provided the possibility of respectably engaging in the art of floristry (or 'floriculture'), a form of flower appreciation associated with the lower classes. Although floristry, and the knowledge derived from it, was usually dismissed in the published work of taxonomic botanists, floriculture was incorporated within the amateur study of botany. The connection was normalised by the cultural association between the study of nature and the visual education necessary for connoisseurship, where significance was placed on cultivating one's taste as well as on cultivating the plants in one's collection.

Writers on botany then drew further connections between their cultivated taste and the practical benefits that it might have for the cultivation of tender exotic plants. They suggested that the amateur of botany's refined vision could enhance his or her practical ability to grow plants. Amateur botany was significant, then, because it valorised an engagement with an effeminate, lower-class form of knowledge that related to florists' flowers. The aesthetically refined, practically orientated, amateur botanist filled the space between the expert collector and the ignorant anthophile.

4   Social Status and the Communication of Knowledge

'The main purpose of Travel... is to learn. To achieve this goal, one must see People and Things.'[1]

The collection and cultivation of exotic flora had a dramatic impact on the content of eighteenth-century botanical collections and on the appearance of European gardens. We have met the plant traders who integrated themselves within the metropolitan scholarly communities of late eighteenth-century London and Paris, and the scholars, amateurs and other consumers with whom they shared and sold specimens and information. The traders did not, however, work alone: a host of hardy plant hunters scoured foreign climes on their behalf, in search of the plants from which new botanical knowledge might be devised, and on which reputations and livelihoods were constructed. Their searches for new specimens led them to penetrate foreign environments and new social worlds; their trajectories cut across political conflicts, demanding forms of expertise that ranged far beyond the strictly botanical.

Like the traders, most plant hunters feature only cursorily in botanical records. As a consequence, they are often referred to as 'suppliers', and seemingly occupied an invisible world, playing an 'ephemeral', although important, role. Their contributions were essential to the further expansion of knowledge about the natural world. As collectors, they brought back the raw materials for scholarly study. They also acted as intermediaries who exchanged information with knowledgeable people dispersed over geographical distance or across the social hierarchy.[2] Detailed evidence about plant hunters is often hard to find in the archives, but historians have begun to tease out information about how such individuals worked, and the roles they played within both the world of science and

---

[1] Memorandum by Horace-Bénédict de Saussure from his personal papers and quoted in Bungener, 'Horace-Bénédict de Saussure', p. 64. My translation.

[2] Schaffer *et al.*, *Brokered World*; Spary, *Utopia's Garden*, p. 69.

the processes of colonisation.[3] Yet, although their significance is now acknowledged, we still know relatively little about how they negotiated the social and cultural contexts that conditioned participation in, and the practice of, eighteenth-century science.

This chapter discusses the journey made by a Scottish gardener, Thomas Blaikie (1750/1[4]–1838), who was sent to the Western Alps to collect plants in 1775. Focusing closely on Blaikie's narrative account, it contributes a micro-historical study of an eighteenth-century scholarly community. We might think of plant hunters as people who simply collected specimens, but their status as go-betweens meant that they gathered – and gave – much more. The chapter assesses the nature of relationships formed between scholars and their associates, the kind of knowledge that circulated among participants, and the differing values attributed to this. Blaikie's narrative leads us to question the social composition of botanical knowledge networks and ultimately to reassess what 'botany' meant in a context where horticulture might be considered just as important as classification and nomenclature.

Thomas Blaikie offers wonderful material for this case study because he kept a detailed journal of his expedition, which, unusually, has survived.[5] Written in phonetic Scots English (and occasionally in phonetic Scots French), Blaikie's diary is not only richly informative of his work as a plant hunter, but also unveils the communities that supported and defined the commerce of people, plants and information, but which have usually been hidden from analysis. Details of such exchanges are rarely preserved: eighteenth-century botanists usually pared down their records to include only information directly related to plants. Blaikie's diary is a singular source for the social history of botany, and provides a unique window on to the challenges that confronted collectors.

---

[3] Murphy, 'Portals of nature'; Parrish, *American Curiosity*; Parsons, *Cultivating a New France*; Raj, *Relocating Modern Science*; Safier, 'Courier between empires', pp. 265–293; Schiebinger, *Plants and Empire*.

[4] Blaikie was born just before Scotland and England synchronised their separate calendars and jointly converted to the Gregorian ('New Style') calendar. According to the Scottish Old Style calendar, he was born on 2 March 1751; over the border in England, however, the same day was 2 March 1750 (in England, the new year started on 25 March). According to the New Style calendar, already in use on the Continent, Blaikie was born on 13 March 1751. Britain converted to the New Style in 1752. See: Parker, *The Earl of Macclesfield's Speech*, p. 5; Taylor, *Thomas Blaikie*, pp. xvi, 5.

[5] The journal has been published in English: Blaikie, *Diary*. There are also two French translations: the Alpine part was translated in 1935, and a full translation has since been published: Blaikie, *Sur les terres*.

In order to gather data successfully, plant hunters had to negotiate social worlds just as much as they had to navigate the natural environment.[6] Thomas Blaikie strode, clambered and scrambled across the Western Alps in search of new specimens. He scaled parts of mountains that had never previously been climbed, crossed vertiginous passes at considerable danger to himself and his guides, and afterwards took respite with cowherds in bucolic Alpine pastures. When he descended to the valleys, however, the young gardener-turned-plant-hunter lived a very different life. Here, Blaikie carefully cultivated scholarly and sociable connections, and tended to the live plants he had gathered. The communities he encountered below the mountaintops tested his skills in sociability just as much as the vegetation and climatic conditions above challenged his botanical knowledge and physical endurance. Creating good relations in the towns and villages nestled in the valleys of the Jura was central to enabling Blaikie to carry out his solitary botanical research among the crags and screes that loomed over them.

Thomas Blaikie's diary thus exposes the social world of early modern natural history travel and collecting. His account of his journey through the Alps invites us to consider how naturalists created connections with one another, how these relationships were structured, and how they were maintained. We would now describe such social connections as a network, and Blaikie's journal also offers us a means of engaging with debates current within the history of science about networks and sociability.[7]

The journal also encourages a reassessment of how knowledge transfer took place, and offers an unusual window on to such processes. As a trained botanical gardener, Blaikie was able to understand and transmit new knowledge between very different contexts. He could comprehend information that was already expressed in conventional botanical terms, but could also access forms of knowledge that were embedded within practices of cultivation and use. Blaikie met and conversed with members of the eponymous Enlightenment network, the Republic of Letters, as well as peasants high up in the mountains. He was able to bridge a conceptual gap between what peasants knew about the natural world and what was known by scholars. Working as a go-between, Blaikie was significant because he could communicate information in both directions.

---

[6] Meredith, 'Friendship and knowledge', p. 154.
[7] The literature on Enlightenment scientific networks is vast, but, for examples, see: Law, 'Methods of long-distance control', pp. 234–263; Livingstone, *Putting Science in its Place*; Lux and Cook, 'Closed circles or open networks?', pp. 179–211; Miller and Reill, *Visions of Empire*; Secord, 'Knowledge in transit'; Spary, *Utopia's Garden*, ch. 2.

Thomas Blaikie

Thomas Blaikie was born in Corstorphine, Edinburgh. His parents ran a small market garden, where he received a basic education in horticulture. Little information remains about his early life, but it is likely that he was educated at the local parish school and then became an apprentice gardener, possibly at the recently reformed Edinburgh Botanic Garden.[8] Blaikie moved to London in the late 1760s, where he probably worked at Kew Gardens, and also at Kennedy and Lee's Vineyard Nursery in Hammersmith.[9] In 1775, he was hired as a plant hunter by two Quaker doctors, John Fothergill (1712–1780) and William Pitcairn (1712–1791), who were amateur gardeners and botanists. They commissioned Blaikie to travel to the Alps to collect rare plants for their gardens. The expedition was a crucial milestone in the young gardener's career, for the plants he brought back to England were living proof of his expertise as a botanical collector. He learnt French and some German, and, significantly, made a series of acquaintances that would prove highly useful to him in later life: an aptitude for languages and a wide range of associates were very desirable qualifications for a botanical gardener in this period. Soon after his return to Britain, Blaikie was hired as gardener to Charles Louis Félicité de Brancas, comte de Lauraguais (1733–1824), a French count and passionate Anglophile. This was a second decisive moment for Blaikie: he emigrated across the Channel and made a name for himself designing fashionable *jardins anglais* for the French aristocracy.

Most of what we know of Blaikie is derived from his journal. Fothergill and Pitcairn required him to keep a daily record of his activities, and he carried this practice on into his later life. Blaikie was affable and apparently not short on charm, but his occasional acerbic comments also show that he was a shrewd social observer. He could be stubborn and opinionated, particularly regarding aesthetics, and was very critical of the gardeners he met who did not 'know plants' botanically.[10] Blaikie

---

[8] Taylor, *Thomas Blaikie*, pp. 8, 11–14. Taylor has undertaken considerable detective work to support these claims, as Blaikie's name does not appear on the lists of pupils at Corstorphine Parish School, and apprenticeship lists are apparently unavailable for gardeners in Edinburgh. However, there is sufficient evidence within Blaikie's diary and from the Blaikie Papers to support these suppositions. The current Botanic Garden at Edinburgh was founded in 1763 by Dr John Hope, amalgamating several existing botanic gardens in the city. On the eighteenth-century history of the Edinburgh Botanic Garden, see: Romero-Passerin, 'Un jardin des lumières'.

[9] Taylor, *Thomas Blaikie*, p. 17. Blaikie possibly also worked in Fothergill's garden at Upton, London. See: Birrell, 'Introduction' to Blakie, *Diary*, p. xii.

[10] Blaikie noted that Mr Williams, a gardener-turned-nurseryman in Paris, 'pretends to do great things but he seems hardly to know any plants'. Blaikie, *Diary*, p. 140. Blaikie's

is also distinguished by his exceptional physical stamina: in the Alps, he regularly walked thirty miles a day, and exhausted almost all of his companions – including one of the future conquerors of Mont Blanc.[11] Blaikie climbed primarily in search of new, unidentified plants. But as his strength increased, it is clear that he was also scaling the mountains simply for sheer pleasure.[12]

Thomas Blaikie arrived in Geneva on 5 May 1775, thirty days after leaving London. Nestled at the westernmost tip of the lake with which it shares a name, Geneva was at this time a tiny French-speaking republic allied with the other Swiss cantons. Its jurisdiction stretched less than six miles beyond its walls, yet Blaikie described with awe its strong fortifications and immense wealth: 'This city', he wrote, 'is supposed to be one of the richest in Europ[e] for its size; there [their] trade is Chiefly in silver and gold work watches and such like'.[13] Popularly renowned for its clockwork, Geneva and its neighbouring cantons were also home to several notable Enlightenment luminaries.

The purpose of Blaikie's journey was to collect alpine plants new to Britain, and he was well rewarded for his efforts: the young gardener received one shilling for every complete species collected (i.e. flowers, seeds, fruits and leaves) and £2. 0s. 10d for every 100 incomplete species. He probably earned around £20 in total for the entire trip, which was equivalent to about a year's pay for a provincial head gardener or an apprentice in James Lee's nursery; he was also promised a premium 'if many species of alpine plants are found in the collection and but few common English ones'.[14] But Blaikie also undertook the commission for non-pecuniary reasons: he stated himself that his 'curiosity' gave him the strength and steely determination to climb ever higher and further in

outspoken response to the Comte d'Artois' failure to appreciate his gardens properly at Maisons in 1777 lost him a lucrative contract to create a new garden there. See: Blaikie, *Diary*, 20 September 1777, pp. 131–132; 26 and 29 November 1777; 6 December 1777, p. 141. See also: Birrell, 'Preface' to Blaikie, *Diary*, p. viii; Taylor, *Thomas Blaikie*, pp. xv–xvi.

[11] This companion was Michel Gabriel Paccard. Blaikie, *Diary*, 2–3 September 1775, p. 78.

[12] For more on the history of mountain climbing, and especially the shift in attitudes towards mountains, see: Hollis, 'Re-thinking mountains'.

[13] Blaikie, *Diary*, 5 May 1775, p. 30. Blaikie usually spelled words phonetically, and I have retained the spellings from Birrell's transcription. Taylor has confirmed that the transcription is as faithful to the original as possible, except that Birrell occasionally added punctuation to clarify Blaikie's meaning. The diary is best read with a Scots accent.

[14] Joseph Banks wrote Blaikie's 'terms of employment' on behalf of Pitcairn and Fothergill. Quoted from Taylor, *Thomas Blaikie*, p. 20. On gardeners' pay, see Chapter 1, p. 43.

search of rare specimens.[15] Most importantly, Blaikie's expedition was significant because it allowed him to develop a reputation as a botanical gardener.

Thomas Blaikie was by no means the first plant hunter to scour the Western Alps. The area is botanically interesting due to the diversity of habitats within the mountainous landscape, which range from dense forests of fir trees to Alpine meadows studded with orchids, and from high lakes to even higher glaciers encircled with rare varieties of anemone and saxifrage. Icy rivers rush through gorges bordered by escarpments upon whose steep sides rhododendrons cling tenaciously. These different locations are very close to one another, yet are characterised by huge variations in temperature and pronounced seasonal differences. They foster an exceptionally varied flora.[16]

Eighteenth-century botanical studies of the area had received a strong fillip from the Swiss botanist and polymath Albrecht von Haller (1708–1777), who had been publishing major works on Swiss flora since the 1740s. In 1768, he published his *Historia Stirpium Indigenarum Helvetiæ Inchoata* (1768), a book that was written in Latin, published in Berne, but dedicated to George III, and which describes and classifies the Swiss flora in remarkable detail. The following year, Haller published a smaller field guide, the *Nomenclator ex Historia Plantarum indigenarum Helvetiæ excerptus* (1769).[17] Blaikie carried this with him during his expedition, and initially occupied himself during rest days by annotating Linnaean binomial nomenclature to his copy 'for the convenience of the Mountains'.[18] Intentionally covering the same area as had several of Haller's assistants, Blaikie botanised across what is now the border between France and Switzerland, from La Ferrière in the north to Chamonix in the south.[19]

Thomas Blaikie's time in the Alps, sketched out in Map 4.1, was divided roughly into four parts. Each expedition demonstrates the significance of sociability to defining knowledge transfer. Between 6 May and 15 July 1775, Blaikie made short trips around Geneva and in the neighbouring mountains of the Jura and Mont Salève (Savoy). He then undertook two lengthier expeditions. The first lasted for almost a month, from 18 July to 22 August: Blaikie walked from Geneva to Aigle, via the Tour Sallière, and then to Berne via the lakes Daubensee and Thunersee. He

---

[15] Blaikie, *Diary*, 24 July 1775, p. 56.   [16] Murith, *Guide du Botaniste* (1810), p. iii.

[17] Haller, *Historia Stirpium* and *Nomenclator* (1769); Sigrist and Bungener, 'First botanical gardens', p. 336; Taylor, *Thomas Blaikie*, p. 52.

[18] Blaikie, *Diary*, 21 May 1775, p. 34.

[19] Ibid., 6–12 August 1775, pp. 68–70; 31 August–6 September 1775, pp. 75–81.

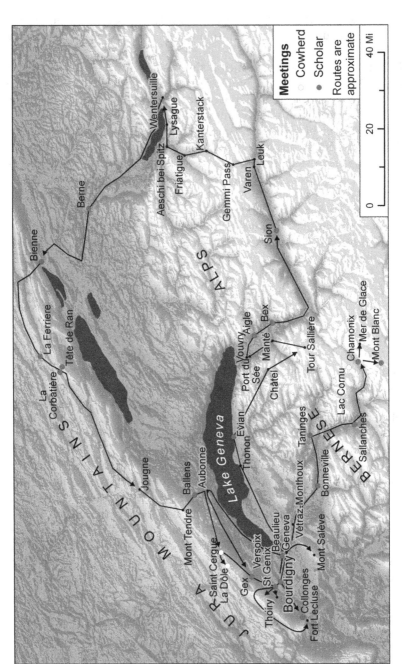

Map 4.1 Thomas Blaikie's itinerary within the Alps.

continued north to La Ferrière, then made his way southwest towards Aubonne via the northern Jura, before finally arriving back in Geneva.

Blaikie's second substantive expedition lasted from 29 August until 12 September 1775. This time, he headed southeast to Chamonix, from where he spent five days scouring the flanks of Mont Blanc for interesting specimens (Mont Blanc, he noted, loomed 'superiore [sic] as a king'[20] over the other mountains). By 5 September, he was ready to box up 'a great number of scarce plants', which he sent back by 'Carryer' to Geneva.[21] Blaikie then set off alone on a weeklong trek across the mountains from Chamonix to Geneva. This expedition in particular was highly perilous: he lost his way and spent a long night shivering among rhododendrons on the mountainside; he was then nearly killed by the xenophobic villagers of Taninges, who believed him to be German.[22] Blaikie escaped, physically unscathed but terribly shaken; for the final portion of his trip, from mid-September until late November, he remained largely within his familiar haunts in the Jura, although he did eventually walk round the whole of Lake Geneva and spent another week collecting near Aigle. He finally left Geneva on 27 November 1775, travelling by stagecoach through France to Britain.

Letters, Introductions and Community

Perhaps surprisingly, given Thomas Blaikie's youth and lack of formal credentials, the gardener visited at least eighteen scholars, meeting the members of a geographically disparate intellectual community. Many, if not all of these scholarly individuals would self-identify as members of the Republic of Letters, a wider, international community of scholars. Its members were united by a mutual commitment to sharing information in order to establish new knowledge.[23] They also enjoyed local affiliations and loyalties to the institutions that hosted them, and to the patrons that supported their research.[24] They communicated through correspondence, via publishing in learned journals and books, and by visiting one another.[25] Poised between the local, the national and the international, the Republic of Letters was fundamental to the exchanges

---

[20] Ibid., 31 August 1775, p. 75.   [21] Ibid., 5–6 September 1775, p. 81.
[22] Ibid., 6–10 September 1775, pp. 81–86.
[23] Brockliss, *Calvet's Web*, p. 4; Jordanova, 'Science and nationhood', pp. 192–211.
[24] On the relationship between the French Republic of Letters and the state, see: Hahn, *Anatomy*, pp. 36–49. For the local institutions and intellectual community in the Western Alps, see: Fuchs, 'Histoire de la botanique en Valais', pp. 119–168.
[25] Goldgar, *Impolite Learning*, pp. 12–34.

that came to define the European Enlightenment. Thomas Blaikie was by no means of a social standing sufficient to be considered a member of this elitist intellectual group, but his journal shows that he attracted its attention and benefited from its assistance. His writings underline the significance of letters of introduction and personal acquaintances to forging links with people who could help him. Indeed, Blaikie's whole itinerary was defined by sociability: he worked out a route that allowed him to visit as many scholars as possible.

Thomas Blaikie's entry into the scholarly circles in the Western Alps was initiated by his British patrons, who arranged that the banker and amateur botanist Paul Gaussen (1720–1806) would act as Blaikie's host and sponsor during the trip.[26] Gaussen cleared space in his own garden at Bourdigny (an estate then just outside Geneva) so that Blaikie could house the new specimens collected; he also arranged the finances for the expedition, managing the money sent by Pitcairn and Fothergill and offering surety so that Blaikie could obtain further funds 'upon the road' if needed.[27] But Gaussen also provided Blaikie with a further, invaluable, service: he introduced the young traveller to other amateurs of botany in the region.

The people to whom Thomas Blaikie was initially introduced lived in and around Geneva, and the introductions took place in person. On 7 May, Gaussen took Blaikie to meet 'a very courious [sic] Gardner' called Mr Martin at the Beaulieu estate.[28] On 22 May, Blaikie 'got aquainted [sic] with Mr Goss [Gosse] a Bookseler [sic] who is exceedingly courious [sic] in Bottany [sic]';[29] the following day, Gosse took him 'to the house of the fameous [sic] professer [sic] of Natural Phylossiphy [sic]', Horace Benedict de Saussure (1740–1799).[30] Three weeks later, Gaussen accompanied Blaikie to Collonges, where he introduced him to Mr Debon, 'a gentellman [who] is very curiouss', who had been working on silkworm cultivation.[31]

---

[26] Paul Gaussen collected a large number of rare trees at his estate at Bourdigny. These included one of the first female specimens of the *Ginkgo bilboa* (maidenhair tree) in Europe. De Candolle, *Histoire de la Botanique Genevoise* (1830), cited in Blaikie, *Diary*, p. 240, n. 1. See also: Sigrist and Bungener, 'First botanical gardens', p. 336; Taylor, *Thomas Blaikie*, p. 24.

[27] Blaikie, *Diary*, 6 May 1775, p. 31, 18 July 1775, pp. 52–53.

[28] Beaulieu was 'a Gentleman's seat' about 1.5 miles from Geneva, 'where there is a hot-thouse [sic] and a Green house'. Ibid., 7 May 1775, p. 32.

[29] Ibid., 22 May 1775, p. 34. The Gosse family had two bookshops in Geneva. Blaikie probably met either Jean Gosse (1727–1805) or his son, Henri-Albert (1753–1816). Ibid., p. 240, n. 2.

[30] Ibid., 23 May 1775, pp. 34–35. On Saussure, see: Bungener, 'Horace-Bénédict de Saussure', pp. 61–66; Sigrist, *H.-B. de Saussure*.

[31] Blaikie, *Diary*, 14 June 1775, p. 42.

But Blaikie's most significant introductions did not, in fact, originate from Paul Gaussen. Instead, they came at second or third hand, via the people to whom Gaussen had introduced him. Further, the introductions were often transmitted by letter rather than in person. This was the case throughout Blaikie's mission, but most especially in Aigle, Berne and Aubonne, towns that Blaikie visited during his two longer expeditions. Blaikie arrived at Aigle, for example, in late July 1775, carrying a letter of introduction from Gaussen to a Mr de Coppet, the town minister (whom Blaikie judged to be a 'verry cevell intelegent man ... [who] seemed to know plants well'[32]). Abraham-Louis de Coppet (1708–1785)[33] adopted the same role in Aigle that Gaussen had performed in Geneva, arranging further introductions. He took his young visitor to a 'friend ... that had been in England and ... [who] knows some plants'.[34] The latter (who is not named in the journal) then introduced Blaikie to one of his most significant acquaintances: 'Mr Decoppets friend ... conducted me to a cottage ... where there lives a very curiouse [sic] man named Abraham Thomas'.[35] Thomas and his son (also Abraham) had collected plants for Albrecht von Haller, and the younger Abraham Thomas (1740–1824) leapt at the opportunity to go botanising with Blaikie: 'we lost no time', the latter wrote, 'but sett of[f] ... to the Mountains.'[36]

Thomas Blaikie and Abraham Thomas junior botanised for three days together, crossing breath-taking Alpine passes, whose magnitude literally stopped Blaikie in his tracks. On seeing the 'tereble' '*Montain* [sic] *de* [sic] *Diable*', a mountain 'cleft in two' by an earthquake, Blaikie 'could not help sitting down' to contemplate the scene. Awestruck, he was even moved to quote from Virgil: 'Dire Earthquakes rent the solid alps below / And from the sumits [sic] shook the eternall [sic] snow.'[37]

Blaikie's ultimate destination was Berne, and the change of language from French to German meant that, to his regret, he was obliged to part company with French-speaking Thomas halfway there. He carried another letter of introduction to Samuel Engel (1702–1784), one of

---

[32] Ibid., 24 July 1775, p. 56.

[33] The pastor Abraham-Louis De Coppet was 'a distinguished botanist, a friend and collaborator of Haller'. Taylor, *Thomas Blaikie*, p. 34.

[34] Blaikie, *Diary*, 25 July 1775, pp. 56–57.

[35] Ibid., 26 July 1775, p. 57. On the significant botanical contributions made by the Thomas family, see: Haller, *Historia Stirpium*, pp. xviii–xxiv; Murith, *Guide du Botaniste* (1810), pp. 4–5.

[36] Blaikie, *Diary*, 26 July 1775, p. 57.

[37] Ibid., 27 June 1775, p. 58. The quotation, which is not attributed by Blaikie, is from Virgil's *Georgics*, Book I (describing the natural portents of Caesar's death).

Berne's most 'distinguished savants'.[38] Just as de Coppet had in Aigle, Engel introduced Blaikie to local amateurs and even took his young visitor 'to see the famous Dr Haller'. The great botanist was, unfortunately, too ill to receive visitors, so Blaikie's new host 'did every thing in his power and gave me a leter [sic] for Mr Gagnebin'.[39] The very next day, Blaikie set off northwards, altering his itinerary in order to meet the physician and amateur botanist Abraham Gagnebin (1707–1800) of La Ferrière, who, according to Engel, 'was little inferiore [sic] to Dr Haller'.[40] Gagnebin, like Blaikie's other new acquaintances, welcomed him warmly and even went into the mountains botanising with him.[41] A week later, Blaikie turned southwards and eventually arrived at Aubonne.

Paul Gaussen's contact at Aubonne, a 'Mr Bonhoit' (i.e. Benoît),[42] responded to Blaikie in exactly the same way as de Coppet and Engel: he showed him local collections and introduced him to other collectors. Most significantly, these included 'a curiouss Gentelleman',[43] Jean-Laurent Garcin (1733–1781), who turned out to be 'a great Botanist', and who, like Abraham Thomas and Abraham Gagnebin, immediately set off botanising with Blaikie.[44] The pair walked fifteen miles to Saint Cergue that same evening, and then spent two nights in the Jura, collecting plants – including some that Blaikie had discovered but that Garcin, although a local, 'had never found'.[45]

It is striking that every recipient of a letter of introduction from Gaussen immediately took their young visitor to meet the local community of amateur botanists. The third- and fourth-hand introductions ultimately led Blaikie to the people best able to help him augment his collections: Abraham Thomas, Abraham Gagnebin and Jean-Laurent Garcin. Those who offered Thomas Blaikie the most significant assistance, and who actually went botanising with him, were very distant from the immediate circle known to his patron Paul Gaussen.

---

[38] Samuel Engel was a Bernese magistrate and author of numerous publications on agriculture, political economy and geography. Blaikie, *Sur les terres*, p. 86, n. 44.

[39] Blaikie, *Diary*, 4 August 1775, pp. 67–68.

[40] Ibid. Dr Abraham Gagnebin was renowned as a botanist, geologist and climatologist. He maintained a sustained correspondence with Haller for over thirty years, and the genus *Gagnebina* is named after him. Ibid., p. 240, n. 4.

[41] Ibid., 6–12 August 1775, pp. 68–70.   [42] Ibid., 13 August 1775, p. 72.

[43] Ibid., 15 August 1775, p. 72.

[44] Jean-Laurent Garcin was seigneur and pastor of Cottens-sur-Aubonne. He was also a poet, botanist and *philosophe*. Blaikie, *Sur les terres*, p. 92, n. 51.

[45] Blaikie, *Diary*, 16 August 1775, p. 73.

Sociability and Scholarly Commerce

Thomas Blaikie travelled with instructions from his patrons in England that detailed what kinds of plants they wished him to collect and the approximate area in which he might search for these. His exact itinerary through the Alps, however, was largely determined by the locations of the people he encountered while travelling, as described in the previous section. The advice they gave him defined his success in finding interesting specimens. Blaikie's experience thus underlines the centrality of sociability to Enlightenment natural history. Blaikie visited a relatively provincial part of Europe, but one in which the local scholarly community was nevertheless well established.[46] His access to knowledge was contingent upon the conventions of sociability, and his journal makes clear the difficulties involved in effective communication, underlining the significance of getting these social interactions right. Blaikie's experiences were markedly different to those of the plant hunters to be discussed in Chapter 5, who travelled to parts of the world as yet barely examined by European scholars. He was further differentiated from many of the scholars he visited (and from the other plant hunters to be discussed) because he was not searching for plants completely new to European science. Blaikie sought specimens new to British gardens. He was guided by a brief that was as much horticultural as it was botanical.

One of the most noteworthy features of Thomas Blaikie's account is the immediacy with which new acquaintances responded to him, even though his arrival was typically unexpected. While such spontaneous hospitality was not unusual according to eighteenth-century conventions surrounding polite visiting,[47] Blaikie's lower social status makes the warm reception he received surprising. Most travellers sporting letters of introduction in Switzerland in the 1770s were on the Grand Tour and hailed from higher social circles.[48] The increasing numbers of such tourists naturally meant a proliferation of letters of introduction, and this consequently resulted in the devaluation of letters as affidavits of merit.[49] Someone of Blaikie's status, furthermore, could not presume to be seen immediately. But many of his hosts rearranged their agendas so that their unexpected visitor would not have to be kept waiting. Samuel Engel spent the afternoon showing Blaikie his collections, and on the

---

[46] Fuchs, 'Histoire de la botanique en Valais', pp. 119–168.

[47] Jones, *Industrial Enlightenment*, pp. 98–99.

[48] The later eighteenth century did, however, see the development of middle-ranking tourism. See: Brewer, *Pleasures of the Imagination*, pp. 631–632, 633; Jarvis, *Romantic Writing*, pp. 16–17.

[49] Chapron, 'Du bon usage des recommandations', pp. 249–250.

following day took him to see a local tourist site.[50] Abraham-Louis De Coppet, who was too old to walk far, devoted a whole day to discussing the plants in his house and garden.[51] Blaikie's experience underlines a developing distinction between 'scientific' travellers and tourists travelling for their own curiosity or Enlightenment.[52]

Blaikie's activities might now be described as 'networking in action', because he rapidly accrued a great number of useful contacts via personal recommendations.[53] Many of his connections, furthermore, were made within an intellectual community for whom making links between members was a major part of its raison d'être. The use of 'network', however, is problematic for eighteenth-century contexts, where the concepts of a 'social network' and of 'network*ing*' had not yet developed.[54] The modern notion of a 'network', which has been described by one sociologist as 'deceptively easy to think',[55] can impose an overly simplified model of power relations and knowledge transfer on situations that were inevitably more complex.[56] As Blaikie's example underlines, information exchange was unpredictable and its effectiveness was defined by factors such as local contexts or the character and astuteness of the individual traveller – factors that are often missed in macro studies of networks.

We have so far seen examples demonstrating successful relationships that Blaikie forged. Each encounter quoted led to support that was

---

[50] Blaikie, *Diary*, 21 May 1775, pp. 35; 3–4 August 1775, pp. 67–68.

[51] Ibid., 25 July 1775, p. 56.

[52] Chapron, 'Du bon usage des recommendations', p. 257.

[53] On networks, start with: Latour, *Reassembling the Social*; Law and Hassard, *Actor Network Theory and After*. For examples of the application of Actor-Network Theory (ANT) to studies of eighteenth-century natural history, start with: Miller, 'Joseph Banks', pp. 21–37; Spary, *Utopia's Garden*, ch. 2. See also: Joyce, 'What is the social', pp. 213–248.

[54] 'Network' in OED Online, www.oed.com/view/Entry/126342 [accessed 11 August 2016]; 'Réseau' in *Le Trésor de la langue française informatisé*, http://atilf.atilf.fr/ [accessed 11 August 2016]. The figurative use of 'network' first appeared in English around 1839 but did not become common until the 1880s. In French, the figurative use of 'réseau' developed slightly earlier, during the first quarter of the nineteenth century, and was then applied to describe social relationships between actors. See also: Mattelart, 'Mapping modernity', pp. 170–174; Spary, *Utopia's Garden*, p. 69.

[55] Law, 'After ANT', p. 6. See also: Latour, 'On recalling ANT', p. 15.

[56] ANT, for example, offers models of knowledge transfer and power accumulation that cannot fully capture historical complexity. Information, according to ANT, flows towards 'centres of calculation' where knowledge is constructed. But the emphasis on a centre–periphery model leaves unanswered the question of how information circulated among 'ephemeral' members of a network, such as Blaikie and his Swiss associates. See: Meredith, 'Friendship and knowledge', pp. 155–156. Some scholars, such as Sairo, *Language and Letters*, have presented eighteenth-century scholarly networks as if they were finite entities composed of clearly enumerable sets of actors. However, networks formed by individuals such as Blaikie were in a continual state of evolution, and it is difficult, if not impossible, to define their extent precisely.

practical (food and shelter) and/or socio-intellectual (conversation, directions to specimens and further introductions). Blaikie and his inter-locutors managed their relationships by conforming to conventions for civility that were widely shared across Enlightenment Europe. This was reflected in the language that Blaikie used to describe his encounters. He measured the social acceptability of his acquaintances according to their behaviour, and described those he approved of as either 'polite' or 'civil'. Paul Gaussen thus 'received me with great politeness'; 'his cevility [sic] was astonishing' and he was soon 'my good friend'.[57] Blaikie kept the same standards in mind when appraising the botanists he met later in the summer. We saw that Abraham-Louis de Coppet was 'verry cevell [and] intelegent'; Blaikie described another acquaintance as 'a man of respect'; others were 'very genteel' and 'very curiouse'.[58] Thomas Blaikie's close observation of the characters of the people he met suggests that, like the plant traders and other gardeners discussed in the previous chapters, the young gardener was in the process of constructing himself as genteel. Politeness, as noted earlier, was central to this construction.[59]

The strength of politeness as a social and cultural mediator meant that its absence – or its breakdown – gained inflated significance. The degree to which transgressions were acceptable was defined both by the local rules of etiquette and by the formality of a specific situation.[60] It was not uncommon for itinerant plant hunters to find themselves negotiating the problem of appearing 'uncivil' to hosts who were inevitably better clad, and often more socially elevated. In the 1730s, Peter Collinson wrote to an American contact regarding the Philadelphian plant hunter John Bartram (1699–1777), whom he hoped would soon pay him a visit. He warned his correspondent, 'Don't be surprised if a downright plain countryman should turn up – you'll not look at the man, but at his mind for my sake ... He comes to visit your parts in search of curiosi-ties.' Collinson also expressed his anxieties about this directly to Bar-tram, writing that 'I do insist, that thou not appear to disgrace thyself or me; these Virginians are a very gentle, well-dressed people, and look perhaps more at a man's outside than his inside. For these and other reasons, pray go very lean, neat, and handsomely dressed.'[61]

For the most part, Blaikie seemed to find his way through the thicket of social conventions. But difficulties did arise occasionally that jeopardised

---

[57] Blaikie, *Diary*, 20 June 1775, p. 45; 6 May 1775, p. 31.
[58] Ibid., 25 July 1775, p. 56; 31 August 1775, p. 75; 26 July 1775, p. 57.
[59] On politeness and gentility, see Introduction, pp. 15–18.
[60] Goffman, *Behavior in Public Places*.
[61] Letters from Peter Collinson to John Custis and to John Bartram, both early 1730s, quoted in Grieve, *Transatlantic Gardening Friendship*, p. 12.

potential information exchange. Cracks in his behaviour began to show, for example, on his more extended encounters with scholars. In early August, Dr Gagnebin, the physician and noted amateur botanist from La Ferrière, took leave for a few days to accompany his new acquaintance into the mountains. The pair initially got on very well: Blaikie had arrived on 6 August, introducing himself 'with the letter I had brought from Berne'; Gagnebin welcomed him and showed him his cabinet of curiosities, and they went botanising in and around the village.[62] On 9 August, they set off into the mountains, staying initially 'at a little village called Ratigue',[63] where they spontaneously joined a wedding celebration: 'we passed with the others the evening very agreeable, the Doctor joined in the dance.'[64] The following day was equally convivial, and they 'returned in the evening to the house we had lodged at last night where we was well entertained as they were the Drs. Aquantance [sic]'.[65] But Blaikie was not able to uphold his genial relations with Gagnebin once they had left the comfort and civilisation of the lower slopes of the mountains. The conditions higher up worsened quickly, and by evening the pair found themselves tense and stressed, lost in a thick fog. '[W]e got to the edge of a steep rock', Blaikie recounted, 'where we could not see the bottom neither did he [Gagnebin] know where he was although he had travelled all over those mountains.'[66] '[T]ired and uneasy', they wandered through the misty gloaming until eventually they 'heard some voices which led us to a little hutt [sic] where the people was with their cattle; here we refreshed ourselves with milk and cheese which was all we could procure and a good bed amongst hay'.[67] Blaikie recalled that 'for me I was very happy', but that this was not so for 'the poor Dr'. Old Gagnebin 'was so mortified to think he had lost his way as he pretended to know so perfectly the road. I could not help laughing at him[,] this hurt his pride so much that he could not sleep.'[68] The tension that had built up during the day broke within the primitive setting of a cowherd's hut. The next day, Gagnebin 'set us right on our way' but a few hours later lost heart and 'desired to return home'.[69] Blaikie's response – laughter – was, of course, not the sole reason why Gagnebin abandoned the expedition. But in Blaikie's narrative of the episode, he stressed that it was *his* behaviour, rather than the inclement environmental conditions or Gagnebin's age (he was sixty-eight) that was at fault. In laughing at his companion, Blaikie had transgressed the bounds of civility and deepened his

---

[62] Blaikie, *Diary*, 6–9 August 1775, pp. 68–69.
[63] I have not been able to locate this village (Blaikie spelt the name as he heard it).
[64] Blaikie, *Diary*, 9 August 1775, p. 69.   [65] Ibid., 10 August 1775, p. 69.
[66] Ibid., 11 August 1775, p. 70.   [67] Ibid.
[68] Ibid.   [69] Ibid., 12 August 1775, p. 70.

companion's existing discomfort. Blaikie's rendition underlines the close attention he paid to monitoring his own comportment; his humiliating outburst of laughter shows the limits, at that point of his career, of his self-presentation as polite.

The conventions that conditioned knowledge transfer were also determined by the social status (and educational level) of interlocutors. This added an element of unpredictability to the process of information transfer, as a plant hunter's success rested both on his botanical expertise and on his ability to get along with the people he met. Notions of politeness offered a framework for Blaikie's behaviour in his dealings with scholars, and their transgression could limit the extent to which information might be exchanged. But judgements about 'civility' or 'politeness' did not extend to those below him in social station. Blaikie befriended cowherds during his travels through the mountains, first encountering those from the Jura in early June 1775, as they ascended the mountains to summer their cattle. Scattered across the Alps, each small community of cowherds provided Blaikie – a stranger – with essential food and shelter during his expeditions. Blaikie appreciated this deeply and actively enjoyed their company: he 'stayed very happy' with 'my good acquaintances' in the Jura and elsewhere.[70]

In contrast, the young gardener repeatedly reported his frustrations with the guides who he paid to lead him from place to place. Not understanding Blaikie's mission, they would refuse to deviate from known footpaths and complained bitterly about the privations that they experienced on the mountainsides. Exploring the Tour Sallière in July, Blaikie described how he and a guide spent their days scaling 'rocks and precepices' and their nights in huts on the mountainside, sleeping on beds made only of hay and subsisting on a diet of milk, cheese, cream and bread. On 22 July, for example,

we returned in the evening to the Hut where we had lodged the night before; here we found no bread and our provisions being finished we was forced to supe [sic] upon cheese and milk without bread which did not please my guide who told me he would go no further with me unless I would promiss [sic] to-morrow to leave the mountains because he could not climb walk and live at that rate and if I entended [sic] to break my neck he would not break his; all this was obliged to agree to and so we laid down upon our hay beds.[71]

Blaikie and his guide descended to the valleys the following day; he secured a replacement, only to find with intense exasperation that 'my guide ... would have continued with me long enough if we had only

---

[70] Ibid., 7 and 10 June 1775, pp. 38–39, 41; 19 August 1775, p. 74.
[71] Ibid., 22 July 1775, p. 54.

gone from house to house or toun to toun but those people having no cureossity [sic] they tire out at once'.[72] Other guides attempted to steal from him, but Blaikie was nevertheless compelled to continue with the same 'Rogues', as he had no other means of finding his way through the mountains.[73] Villagers likewise often treated the young traveller with hostility or even fear, refusing shelter or food, and in one case physically attacking him.[74] The cowherds' openness was unusual, therefore. Information, rather than civility, was the key mediator through which Blaikie forged connections and bridged differences. In this case, *what* one knew (and what one would share) mattered most.

The rules of civility that smoothed relations among members of the Republic of Letters did not apply to either the cowherds or the guides. Neither was described as 'polite', for example, although the former were 'very hospitable'. Blaikie's modified language here signals his sense of distinction over and above both groups, but the perceived lack of politeness was not necessarily a criticism and did not hamper the exchange of information, where information was to be exchanged. The young gardener enthusiastically described how his 'friends' the cowherds were always willing to share their meagre provisions with him, and would press him to stay longer with them on the mountainside.[75] In the pages of his diary, Blaikie represented his bucolic hosts almost romantically, as peasants whose simplicity and connection to the natural world positively distinguished them both from his more polite, urbane associates and from the guides who hindered rather than helped his mission.

Thomas Blaikie, a twenty-four-year-old gardener from the outskirts of Edinburgh, met and learnt from a wide range of scholarly amateurs in the valleys of the Western Alps. He also enjoyed and profited from the company of the cowherds in the mountains overshadowing them. These peasants offered him much more than convenient and welcoming places of refuge, providing a service that professional guides could not offer: Blaikie recorded that they conducted him to locations where he was most likely to find good specimens. Knowledge transfer depended on the successful creation of relationships across the social hierarchy.

Defining the Pathways of Knowledge

The warm reception that Thomas Blaikie received from the scholars on whom he called is understandable given that he performed a subtle but

---

[72] Ibid., 24 July 1775, p. 56.   [73] Ibid., 30 July 1775, pp. 62–64.
[74] Ibid., 8–9 September 1774, pp. 84–86.
[75] Ibid., 7 September 1775, p. 84; 3 and 4 July 1775, p. 49.

important function as an intermediary between his local hosts. The circulation of letters of introduction, discussed earlier, served to strengthen the existing connections among scholars. As material objects, the letters symbolically manifested the continuing friendship between writer and recipient, functioning as forms of 'moral currency', particularly if the bearer was someone of potential value to the new host.[76] From the perspective of Enlightenment sociability, it was important to meet a new acquaintance face to face: such encounters could count very highly for establishing links that might be called upon later.[77] But the gardener Thomas Blaikie could not offer his new acquaintances anything of social significance. Blaikie's contribution related directly to the information that he carried with him.

The connections that Thomas Blaikie made were grounded on a notion of reciprocity in which knowledge about plants was the primary commodity. Blaikie was received warmly because he could communicate something new and interesting to his hosts. This, of course, differentiated him from most other tourists. During the course of his travels, Blaikie forged connections with a range of 'knowledge communities', as already discussed. He was consequently able to access and transmit information that formerly had been exclusive to one group or another. As an intermediary, Thomas Blaikie defined what was transferred, and between whom.

The significance that Blaikie placed on conversation in his written accounts of his days parallels that expressed by the metropolitan gardeners and plant traders. In their case, this extended from conversation to participation in correspondence networks, through which information was circulated in written form.[78] Indeed, the nurseries that enjoyed the greatest commercial success were those that were run by the most assiduous and reliable letter writers.[79] It should come as no surprise that Andrieux and Vilmorin promised to reply to all correspondence, and that James Lee was equally diligent.

Gardeners who failed to live up to expectations for communication, such as William Aiton (1731–1793), head gardener of the royal gardens

---

[76] Jones, *Industrial Enlightenment*, p. 98; Meredith, 'Friendship and knowledge', pp. 157–158. See also: Ogborn, 'Writing travels', p. 158 and Chapron, 'Du bon usage des recommendations', pp. 249–258.

[77] Hancock, 'The trouble with networks', esp. pp. 477–483.

[78] Brockliss, *Calvet's Web*, pp. 79–104; Goldgar, *Impolite Learning*, ch. 1.

[79] According to Silvestre, Vilmorin's letterbooks were considered essential resources for members of the Société d'Agriculture. It is a great shame that so little of the Andrieux and Vilmorin correspondence has survived. See: Silvestre, 'Notice biographique', p. 199.

at Kew, were strongly criticised.[80] Aiton was neither lazy nor unsociable: his acquaintances recalled pleasant face-to-face meetings with him and complimented him for his intelligence and skill. One of Joseph Banks' French correspondents remembered how, during his visit to England in the late 1780s, 'Mr. Aiton took me to see the vegetable collection at Kew with all possible generosity and civility, and all the pleasure that an excellent botanist like him senses in showing rare and curious plants. I shared his enthusiasm well, and in appreciating his merit, I took back to France the highest idea of his talents.'[81] Aiton's friends, however, were surprised to hear that he 'does not like to communicate'.[82] The Swiss botanist and professor of medicine, Johann Jacob Römer (1763–1819), wrote to Joseph Banks in frustration because, although Banks had encouraged him to send a box of desirable seeds as a present to Aiton, 'I have not received one word of response: far from it that Mr. A. had sent me some seeds or plants in exchange, he has not even bothered to let me know that he had received the seeds. I therefore sent him a short letter in English, in which I asked him to reply, ... but on this letter also I have remained without response.'[83] Aiton's behaviour caused disquiet within the wider community. The amateur of botany Sir Richard Kaye (1737–1809) recorded a conversation between several botanical friends during a visit to Kew, in which they noted that Aiton's failure to communicate meant that he 'loses plants' and 'is often obliged to come to these, [via collectors] who have been supplied elsewhere'.[84]

Traders of natural history specimens depended on maintaining good relationships with their correspondents. Emanuel Mendes da Costa, the London-based natural history trader discussed previously, conducted an extensive correspondence with his customers and other commercial contacts.[85] Like Lee and Vilmorin, da Costa used letters to establish and maintain relationships with customers, and to obtain specimens from travellers and other vendors. He was initially an assiduous correspondent: lapses were unusual and subject to jokes rather than serious criticism. After an atypical silence in 1748, Peter Collinson postulated that 'either Love had turn'd his head, or [he is] Lost in some

---

80 Drayton, *Nature's Government*, pp. 46–48, 125–127.

81 BL, Add. Ms. 8097, ff. 162–163, Baron Dumont de Courset (Château de Courset, Boulogne sur Mer) to Joseph Banks (London), 22 July 1789.

82 BL, Add. Ms. 18565, Kaye Notebooks, Vol. XVI, f. 83r.

83 BL, Add. Ms. 8098, f. 282, Römer (Zurich) to Joseph Banks, 28 March 1793.

84 BL, Add. Ms. 18565, Kaye Notebooks, Vol. XVI, f. 83r. Kaye did not date this conversation, but it probably took place around 1790.

85 Cantor, 'Emanuel Mendes da Costa', pp. 230–236; Rousseau and Haycock, 'The Jew of Crane Court'.

mine... purloining of Fossils'.[86] But da Costa became less punctual as he grew older: he complained that his days were filled to excess with trading, travelling, collecting and publishing on natural history, and it could take him several months to reply to letters.[87] As with Aiton, this annoyed his correspondents: in 1764, Peter Simon Pallas (1741–1811) tetchily told da Costa that he had lost a potential contact who 'did not like irregular correspondents; for, give me leave to tell you, this is the character, you generally bear abroad'.[88] Da Costa nevertheless understood that letter writing was very important. He made careful copies of his own letters, prefaced many of them with effusive apologies for his delay, and assured his correspondents that he would always (eventually) send them replies.[89]

Information about the natural world can, of course, be transmitted in physical objects as well as via the written or spoken word. In contrast to the other natural sciences, particularly zoology, it is relatively easy to codify information about plants, and thus to transfer that information across distances. Plants could be represented visually in easily transportable formats such as herbarium specimens or botanical paintings that followed precise representational conventions. Both were usually accompanied with brief textual descriptions that, thanks to innovations such as Linnaean binomial nomenclature, worked as shorthand for lengthier verbal descriptions. These physical objects, described by historians of science as 'immutable mobiles' (using a phrase coined by Bruno Latour), could convey information that was difficult to capture succinctly in words.[90] They had to be preserved and made portable according to specific scientific conventions, to ensure that they could be incorporated into existing collections, and so studied and compared with ease.[91] One

---

[86] BL, Add. Ms. 28536, f. 60r, Peter Collinson (London) to da Costa, 26 September 1748.

[87] Rousseau and Haycock, 'The Jew of Crane Court', p. 137.

[88] BL, Add. Ms. 28540, f. 279, Peter Simon Pallas to da Costa, 10 November 1764. Also cited in Rousseau and Haycock, 'The Jew of Crane Court', p. 137.

[89] Letterbooks containing drafts and copies of da Costa's extensive correspondence are held in the British Library, classmarks Add. MSS 9389, 28534, 28544, 29867, 29868, Egerton Ms. 2381. Further correspondence can be found in Derby Local Studies Library, the Fitzwilliam Museum, Cambridge and Warwick County Record Office. See: WCRO, CR2017/TP408, f. 11, Draft letter from Emanuel Mendes da Costa (London) to Thomas Pennant (Downing), 27 June 1752; WCRO, CR2017/TP408, f. 17, Draft letter from Emanuel Mendes da Costa (London) to Thomas Pennant (Downing), 1 September 1752.

[90] Latour, 'Visualization and cognition', pp. 1–40 and *Science in Action*, pp. 227–237.

[91] Latour's concept of immutable mobiles has been discussed extensively by historians of science. See, for example: Bleichmar, *Visible Empire*, pp. 60–65, esp. p. 63; Golinski, *Making Natural Knowledge*, pp. 98–99; Miller, 'Joseph Banks', pp. 21–37; Spary,

preoccupation of Enlightenment botanists, consequently, was with per-
fecting methods for ensuring the survival of specimens transferred as
seeds or as growing plants (the latter were very complicated to transport,
as Chapter 5 will explain).[92]

Thomas Blaikie put his botanical and horticultural expertise to use by
carefully preserving his Alpine collections according to these established
conventions. He collected seeds and sent a huge number of live plants
to Britain.[93] He also closely studied the botanical gardens that he visited
during his expedition, even making his own catalogue of the plants in the
Jardin du Roi in Paris, and he preserved plants for inclusion in herbaria
(Figure 4.1).[94] The seeds, herbarium specimens and live plants that he
collected served, in effect, as means by which the Alpine world through
which he travelled could be 'abbreviated' and 'flattened'.[95]

The form in which information is represented affects the extent to
which knowledge can be transferred. Gardens offer a further means of
rearticulating natural knowledge, and, although ostensibly sites of complete immobility, they have also been essential in facilitating longer-
distance plant transfers. Thomas Blaikie created two gardens in which
he housed temporarily the live specimens he collected from the mountains. One was in Paul Gaussen's estate at Bourdigny, and the other was
in the small village of Saint-Genis-en-Pouilly, which was closer to the
Jura.[96] The value that local scholars attributed to his collection, particularly the garden at Saint-Genis, was clear: several luminaries went to
visit it independently, including an aging Voltaire, who, Blaikie noted
with pride, was 'very well pleased with my collection'.[97] Seven weeks
later, when Blaikie was introduced to Jean-Laurent Garcin, he learnt to

---

Utopia's Garden, pp. 84–85, 97; Stemerding, Plants, Animals and Formulae, pp. 85–89;
Withers, Geography, Science and National Identity, pp. 20–22.

[92] For more on this, start with: Easterby-Smith, 'Reputation in a box'; Parsons and Murphy, 'Ecosystems under sail', pp. 503–529.

[93] Blaikie brought 440 species of plants back to Britain (amounting to over 3500 specimens in total). Blaikie's plant list has survived and is reproduced as appendices to both
published editions of Blaikie's diary, as well as in Taylor, Thomas Blaikie, pp. 52–53. See
also: Birrell, 'Introduction' to Blaikie, Diary, p. xi.

[94] Figure 4.1 shows the only known example of a herbarium specimen with Blaikie's
name on it. Blaikie introduced thirty-two or thirty-three new Alpine plants, but they
are attributed to Pitcairn and Fothergill. The 1789 edition of William Aiton's Hortus
Kewensis lists four plants as introduced by 'Mr. Thomas Blackie', but these introductions were made in 1777 and 1778, once Blaikie's contract with Pitcairn and Fothergill
had terminated, and were presumably not part of the Alpine collection that he had sent
to his patrons. Aiton, Hortus Kewensis (1789), vol. I, pp. 80, 144; vol. II, p. 285; vol.
III, pp. 423–424. See also: Clifton, 'The Blaikie list', pp. 1–50; Taylor, Thomas Blaikie,
pp. 52–53.

[95] The words quoted are from the article 'Histoire naturelle' in the Encyclopédie, discussed
in Stemerding, Plants, Animals and Formulae, p. 87.

[96] Blaikie, Diary, 9 May 1775, p. 32.   [97] Ibid., 24 June 1775, p. 47.

Figure 4.1 *Senecio alpinus*.[98] Collected by Thomas Blaikie in 1776. By permission of the Linnaean Society of London.

[98] The plant's current botanical name is *Senecio subalpinus* (Koch). Using Haller's *Nomenclator*, Blaikie identified it as *Cineraria cordifolia* ('Heart-leaved Cineraria'). It was recorded as such in the *Hortus Kewensis*, which states that it was 'Introd[uced]' in 1775 by the Doctors Pitcairn and Fothergill'. Aiton, *Hortus Kewensis*, vol. III, p. 221. See also: Taylor, *Thomas Blaikie*, p. 229.

his great surprise that Garcin had already travelled the distance of thirty-five miles 'to see my collection of plants'. Garcin too was apparently 'well pleased' with what he saw there.[99] Blaikie may have been working for his British patrons, but his collection nevertheless demonstrated to local scholars his own botanical proficiency.

The significance placed on verbal communication (either face-to-face or via letter) qualifies the Latourian emphasis on immutable mobiles. Visual and verbal representations can only communicate part of the necessary information about flora. Plant hunters such as Blaikie potentially contributed much more than simply gathering together and preserving specimens in collections for others to study, and this is why it was important for scholars to meet them in person. Blaikie and his counterparts were important to botany because they developed specific forms of expertise as a result of their practical knowledge and their itinerancy. Thanks to his contact with peasant communities in the mountains, Blaikie gained unique access to information that had not yet been fully articulated as scientific knowledge, and certainly not codified in ways comprehensible to botanists.

In Chapter 3, we saw how the culture of amateurship encouraged – and justified – an upper-ranking engagement with the practical knowledge required for the cultivation of ornamental flowers. The association of such plants with effeminacy and with the culture of the lower ranks, however, meant that such knowledge otherwise held a low status. In great contrast to the enhancement of the appearance of otiose florists' flowers, improving the cultivation of specimens used in agriculture, industry or medicine gained pressing importance in an Enlightenment Europe concerned with supporting and sustaining an apparently expanding population.[100] The values attributed to *savoir-faire* depended greatly on the plants in question.

Scholars concerned with agricultural improvement thus showed increasing interest in peasants' knowledge during the Enlightenment, seeking to understand the practical expertise already developed among, in our case, the cowherds.[101] However, although developed closer to home, the forms of expertise cultivated within peasant communities

---

[99] Ibid., 15 August 1775, p. 72.

[100] Enlightenment savants' attitudes towards agricultural improvement and the use of natural resources varied greatly. At stake were different judgements about the potential that agriculture might have for overcoming perceived 'natural' limits. These debates are examined by Fredrik Albritton Jonsson and Peter Jones in, respectively, *Enlightenment's Frontier* and *Agricultural Enlightenment*. See also: Gascoigne, *Science in the Service of Empire*, pp. 147–151.

[101] Bourde, *Agronomie et Agronomes*, vol. 1, pp. 12–20; Bungener, 'La botanique', pp. 290–294; Jones, *Agricultural Enlightenment*, ch. 3. For a helpful discussion of the limitations

could appear to scholars just as foreign as those emerging from European colonies. At heart, the issue related to communication, especially given the emergence of increasingly specialist vocabularies within intellectual communities. Interest was directed both at the techniques peasants used and also at their understanding of plant physiology and nomenclature. In a letter to Philippe-Victoire Lévêque de Vilmorin of 1792, for example, the Abbé de Paix noted a distinction between local people's ability to name a plant and their ability to cultivate it. At this time, Vilmorin was researching types of pine tree, and his correspondent noted that, '[t]he cultivators of this country do not know [connaître] the difference between the pine and the fir tree. They confuse the two genera in denomination, but not in cultivation.'[102] The type of knowledge required to name a plant is not the same as that required to cultivate it: one is 'propositional' (knowing 'what'), the other 'prescriptive'·(practical knowledge; knowing 'how').[103] De Paix observed that even though cultivators did not consistently differentiate between the trees by name, they nevertheless perceived their differences and nurtured them appropriately. Their observation of variations between plants was grounded in a practical savoir-faire that did not fit into received botanical discourse.

The local cultivators' perceptions of the natural world were evidently so different to those of de Paix that they were unable to find words with which to explain to him what they knew. Thomas Blaikie experienced similar difficulties communicating in the Jura. He discussed the local flora with the cowherds, noting that 'severals [sic] . . . knowed [sic] many of there [their] plants according to the names of the country as likewise the uses they aplyed [sic] them for either for themselves or for there cattle'.[104] He found that they 'frequently was very carefull [sic] to explain to me as likewise to ask of me what I thought of them [the plants] and what was there [their] uses'. But, Blaikie noted, the cowherds struggled to communicate with him: 'frequently where they had observed any remarkable plant which struck there [their] fancy they would conduct me . . . when they found they could not make me understand by there [sic] descriptions; this happened to me frequently.'[105]

This incommensurability was similar to that which the Abbé de Paix had observed, and in both cases relates to the problem of communicating knowledge that is 'implicit'. Implicit knowledge describes something that a person knows but is unable to communicate – often because they

---

of the Encyclopédistes' engagement with useful knowledge, see: Roberts, 'Circulation of knowledge', pp. 47–68.

[102] AN, 399/AP/101, Chartrier de Malesherbes, Copy of letter from M. l'abbé de Paix (Chanoine de Liège) to Vilmorin, 26 April 1792.

[103] Mokyr, Gifts of Athena, pp. 4–5.   [104] Blaikie, Diary, 3 July 1775, p. 49.

[105] Ibid.

lack the vocabulary to explain how they know what they do.[106] As it has not been articulated, this knowledge remains tacit, and is best learnt by travelling to the field. Blaikie's ability to access the implicit, uncodified knowledge of peasants, in addition to the rare specimens he collected, made it important for the local scholars to meet him in person. Gardeners such as Blaikie were thus differentiated from the more sedentary and bookish savants.

Many of the people who encountered Thomas Blaikie in the valleys appreciated their conversations with him very deeply. Blaikie was valuable because he could communicate information about the Alpine flora that ranged from the botanical to the horticultural, and which included a large degree of practical knowledge. From a horticultural perspective, this included observations about the habitats best suited to cultivating specimens – information that was essential for anyone seeking to grow Alpine plants in gardens. Blaikie saved the nurseries back home in Britain much trial and error.[107] He was very attuned to the beauty of the landscape, and frequently recorded his aesthetic observations. On his journey to Chamonix, for example, he noted how 'as I advanced the country is more mountaineous [sic] and in several places most beautifull [sic], cascades falling from those rocks which are of a prodegeous [sic] height and where the road serpents through this narrow opening formes [sic] at every step a differant [sic] Landskape [sic] beyond all emagination [sic]; here there is a great many curious plants.'[108] Blaikie's Alpine experience helped him to refine his own aesthetic sensibility; this later enabled him to become one of the foremost garden designers in France, creating *jardins anglais* for some of the country's highest aristocrats.

As suggested by the quotation at the start of this chapter, the principal aim of Enlightenment travelling was to learn, and learning necessitated meeting people and seeing things. Many historians have underlined the significance of travel in the construction of naturalists' correspondence networks, emphasising the centrality of face-to-face encounters to creating connections that then developed into corresponding relationships.[109] For Thomas Blaikie, however, botanical travel was not simply a matter of instituting and extending exchanges with other 'scholars'.

---

[106] Knowledge of the rules of generative grammar is an example of an implicit form of knowing. For more on the distinctions between explicit, implicit and tacit knowledge, see: Davies, 'Knowledge (explicit and implicit)'.

[107] See, for example, Blaikie's description of the plants growing on the flanks of Mont Blanc. Blaikie, *Diary*, 31 August–2 September 1775, pp. 74–78.

[108] Ibid., 30 August 1775, pp. 74–75.

[109] Hodacs, 'Linnaeans outdoors', pp. 183–209; Lux and Cook, 'Closed circles or open networks?', pp. 183–191; Meredith, 'Friendship and knowledge', pp. 162–163, 189–190.

Itinerant go-betweens such as Blaikie have been central in the formation of connections between 'disparate worlds', defining what have been
termed the 'pathways of knowledge'.[110] Blaikie's journal has exposed
something of what plant hunters contributed to those pathways: his
hosts recognised him as someone who possessed a good knowledge
of botany coupled with practical skills essential for plants' cultivation.
Social astuteness and adaptability were further essential skills. Blaikie
was relatively unusual because he was able to move among diverse knowledgeable groups and could access information that was tacit or implicit,
as well as that which was already codified.

Blaikie forged new connections with scholars in the Western Alps;
these links benefited him in the short term, because his new acquaintances could offer him practical assistance and could tell him about
good locations from which to gather rare specimens. He also maintained contact with several of these acquaintances, exchanging specimens and information with them over the ensuing decades. It is
important to remember, too, that Blaikie also influenced the *existing* relationships within that intellectual community. Local scholars implicitly
reaffirmed their links with one another by offering Blaikie further introductions within their community. We see here, then, a double movement: on the one hand, Blaikie's future career was forged through the
acquaintances he made; on the other, he also contributed to reinforcing relationships already established among Swiss scholars. Although
Thomas Blaikie did not consciously consider it in this way, a form of
'networking' (in the modern sense) was enfolded within the logic of his
journey.

Blaikie's web of connections, furthermore, was a protean entity. Most
of its members adhered to logics of sociability that were further refracted
through social status: Blaikie was assisted (and occasionally impeded) by
individuals who ranged from scholars to peasants, and from innkeepers
to local guides and couriers. Most of these encounters were relatively
transitory, but each was significant at the time. Enumerating his contacts
tells us little; understanding the strength and nature of the relationships
formed is much more significant.[111] After all, the extent to which Blaikie
was able to enlist help from these groups defined his achievements during
his mission.

---

[110] Schaffer *et al.*, *The Brokered World*, pp. x–xiv.
[111] Granovetter, 'Strength of weak ties', pp. 1369–1373, 1377–1378 and 'Economic action
and social structure', pp. 481–510; Lux and Cook, 'Closed circles or open networks?',
pp. 181–182.

The members of the scholarly communities that composed the Republic of Letters are usually considered to have been responsible for Enlightenment knowledge-making. Thomas Blaikie, however, presents an alternative picture. His example has underlined that knowledge transfer depended on crafting personal relations with a wide range of informants, and thus on the capacity of each individual traveller to comprehend new data and to communicate it to others. As an intermediary, he was uniquely placed to select and transfer data not available to his more elite counterparts. He then translated that information into the vocabulary used by his intellectual associates. Social status mattered, with regards to both access to knowledge and its subsequent communication.

5   Commerce and Cosmopolitanism

The new plants acquired by British and French nurseries and botanical institutions potentially served loftier concerns than those relating to either curiosity-led scholarship or the profitmaking desires of the commercial nurseries – both evoked in the previous chapters. This is evident particularly in the case of the collection and study of exotic plants, which, although they certainly enriched horticultural planting schemes, could also serve much more utilitarian ends, bolstering European economic and imperial agendas. Empire looms large within eighteenth-century initiatives to collect biota from the wider world.

Viewed from this perspective, botanical collecting loses any gloss of neutrality. The collection and study of plants served imperial schemes in several ways, offering practical information that might ensure colonists' survival, revealing resources for economic exploitation and, towards the end of the eighteenth century, even offering a moral justification for the colonial appropriation of land and natural resources.[1]

Colonial botany in the eighteenth century did not necessarily carry with it an agenda of territorial appropriation. Gathering information, especially about natural resources, was often considered more valuable than the land-grabbing that eventually characterised nineteenth-century imperialism.[2] Aiming to revitalise and enhance agriculture in Sweden, Carl Linnaeus placed priority on the importation of economically useful specimens that might be acclimatised and then exploited within his native kingdom.[3] Substantive national collections of rare and lucrative plants, further, served a symbolic role as projections of imperial power. In 1787, Joseph Banks proposed to convert Kew Gardens into a 'great

---

[1] Drayton, 'À l'école des français', p. 95 and *Nature's Government*; Safier, *Measuring the New World*; Schiebinger, *Plants and Empire*; Schiebinger and Swan, *Colonial Botany*.

[2] Bleichmar, *Visible Empire*, ch. 1; Spary, 'Peaches which the patriarchs lacked', pp. 14–41. See also: Delbourgo and Dew, *Science and Empire* and Bayly, *Empire and Information*, esp. pp. 271–275.

[3] Drayton, *Nature's Government*, pp. 72–73.

centre of plant exchange for the Empire', at which plants might be analysed and then sent out for economic exploitation in British colonies. The Gardens themselves were considered to represent the world in microcosm.[4] By the latter years of the eighteenth century, the symbolic function performed by such gardens was particularly important for the assertion of claims over areas that had not been formally colonised, especially in the Pacific.[5]

Eighteenth-century botany took place, then, within a context of national rivalries over natural resources and information that might serve imperialistic ends. Yet, as this book argues, many of the Europeans who gathered prized information and valuable specimens primarily identified with notions of amateur scholarship. The Republic of Letters facilitated international exchange between amateurs; its members formed relations with one another in accordance with its social structure and codes of comportment.[6] The ideas and practices so generated are usually characterised as 'cosmopolitan', because their main features were the impartial exchange of information and specimens.[7]

Behaving as a cosmopolitan meant putting aside specific loyalties in favour of sharing knowledge and information for the greater good of humanity.[8] Fed by the conviction that wisdom could be achieved only through the unconstrained exchange of information, cosmopolitanism demanded political dualism and encouraged international scholarly cooperation.[9] This apparent commitment to impartiality, however, throws into question the activities of both scholarly collectors and plant traders, for both had reason to restrict the international flow of information. Collectors might obtain 'useful' plants that could potentially serve national and imperial agendas. Commercial nurseries, seeking to make

---

[4] Drayton, 'À l'école des français', pp. 100, 106–107. For the parallel Spanish case, see: Bleichmar, *Visible Empire*, ch. 4.

[5] Drayton, *Nature's Government*, chs 2 and 3; Lacour, 'La place des colonies', pp. 49–73, esp. p. 72; Lipkowitz, 'Seized natural history collections', p. 20; Starbuck, *Baudin*, p. 4.

[6] Brockliss, *Calvet's Web*, pp. 8–13; Goldgar, *Impolite Learning*. National and regional academies offered local institutional bases for the Republic of Letters and gave a local inflection to an international movement. See: Hahn, *Anatomy*, ch. 2; Lipkowitz, 'Seized natural history collections', p. 20.

[7] Note that the term 'cosmopolitan*ism*' is an anachronism for this period. It developed in the nineteenth century in response to the emergence of nationalism. I use it to denote the general cultural phenomenon, rather than a specific ideology.

[8] This definition contrasts to the common modern understanding of 'cosmopolitan' as a synonym for 'international'. For more on the shifting meanings of this term, start with: Goldgar, *Impolite Learning*, pp. 174–218, esp. 194, 202; Kleingeld, 'Six varieties of cosmopolitanism', p. 507; Schlereth, *The Cosmopolitan Ideal*, pp. xvii–xxv.

[9] Schlereth, *The Cosmopolitan Ideal*, Preface.

a financial profit from rare plants, occupied an even more contradictory position. In both cases, we might wonder whether they were really willing to exchange facts and specimens impartially.

This chapter and the one that follows examine the evolution of cosmopolitan attitudes from the perspectives of both the 'official' scientific communities and the commercial ones that acted as pendants to them. Commercial plant traders and members of the Republic of Letters shared the same global networks through which plants were conveyed. The cosmopolitan ideal was attractive to individuals who sought to make a financial profit from science, as well as to those who pursued science for the sake of curiosity. The plant traders used this ideal as a cultural formulation, which they adopted and promoted as and when it suited them. Following the contours of the 'age of revolutions', a period that stretched from the 1770s to the early nineteenth century, this chapter examines the impact that the American War of Independence (1775–1783) and the French Revolutionary and Napoleonic Wars (1792–1815) had on cosmopolitan plant exchanges. Chapter 6 then focuses specifically on how the substantive social and ideological changes ushered in by the French Revolution affected the fortunes of French plant traders and their international counterparts. Wars and revolutions jeopardised not only the international exchange of plants but also the commitment to cosmopolitan impartiality overall. Were the bonds that held the Enlightenment scholarly community together brutally severed in times of political conflict and heightened nationalism?

Putting an Ideal into Practice

The practices associated with collecting and correspondence ostensibly supported scholarly cosmopolitanism: possession of some type of collection was a key characteristic defining scholars' identities; cabinets were stocked with items obtained, variously, from travels abroad, from exchanges with other scholars, or by purchase from traders.[10] But a tension existed within this apparently innocent pastime: the collection of naturalia, including plant specimens, had long been encouraged for imperialistic and economic reasons as well as for private edification. Scholars had regularly issued instructions to travellers explaining how to collect and bring back the curiosities that they encountered. By the

[10] Brockliss, *Calvet's Web*, pp. 14, 87–88, 91; Laissus, 'Les cabinets d'histoire naturelle', pp. 659–660, 664–666; Pomian, *Collectionneurs*, pp. 163–194; Swann, *Curiosities and Texts*, pp. 1–4, 194–198.

later eighteenth century, however, these instructions to travellers were increasingly developed under the auspices of the state.[11]

The upper-end plant traders, like scholars, obtained plants through maintaining personal connections with overseas voyagers. Hiring individual collectors was extremely expensive: Kennedy and Lee claimed in 1815 that over the preceding seventy years, their nursery had sent men to collect for them in the Cape of Good Hope, North America and China, but the nursery was not actually the sole patrons of these people.[12] The solution, which was widely practised by the elite nurseries, was to become integrated within the networks of collectors that structured and supported the Republic of Letters. Andrieux and Vilmorin maintained close relationships with plant hunters commissioned by other patrons or institutions. André Michaux (1746–1802), for example, officially collected for Louis XVI and the Jardin du Roi, but also regularly sent North American plants to a range of other Parisian patrons and contacts, including Andrieux and Vilmorin.[13] Lee and Kennedy united with Joseph Banks in aid of Francis Masson's (1741–1805) global collecting expeditions, and with Banks and John Fothergill to finance William Brass' (d. 1783) mission to the Gold Coast and the Ashanti Country.[14] In most cases, the late eighteenth-century nurseries probably did not give substantive financial support to these collectors. Their contributions were conceived more as an exchange of information and as services between scholarly friends.

French collecting networks operated slightly differently compared to British, because of the dominance of the Académie Royale des Sciences. The close ties between the Académie and the government resulted in the earlier development of crown-supported initiatives to collect, study and transfer plants between French possessions.[15] Collectors overseas

[11] For examples, see: Barrin de La Galissonière and Duhamel du Monceau, *Avis pour le transport par mer des arbres* (1753); Ellis, *Directions for Bringing Over Seeds and Plants* (1770). For critical discussion of the instructions issued to collectors, start with: Allain, *Voyages et survie des plantes*; Kury, 'Les instructions de voyage', pp. 65–91; Mackay, *In the Wake of Cook*, pp. 14–15; Parsons and Murphy, 'Ecosystems under sail', esp. pp. 504, 510, 534–526; Spary, *Utopia's Garden*, pp. 78, 82–88; Swan, 'Collecting naturalia'.

[12] NHM, H.B. Carter Transcript, James Lee (Hammersmith) to Joseph Banks, 30 August 1815.

[13] Heuzé, *Les Vilmorin*, p. 12; Robbins and Howson, 'André Michaux's New Jersey garden', pp. 351–370; Silvestre, 'Notice biographique', p. 195.

[14] Willson, *James Lee*, pp. 42–43; NHM, Banks Collection, 'Receipt for 5 guineas received from Banks and paid to James Lee by William Perrin for 42 species of seeds sent home by William Brass.' See also: Carter, *Sir Joseph Banks*, p. 174.

[15] Drayton, 'A l'école des français', esp. pp. 102–103.

enjoyed, furthermore, a choice range of diplomatic, military and commercial channels through which they could send back specimens. Naturalia was consequently often transferred to France more efficiently than to Britain. British collectors were forced to negotiate less reliable ad hoc arrangements, usually with the merchant navy.[16]

We have already seen that plant hunters depended heavily on personal contacts created through travel, visiting and correspondence. Indeed, correspondence was essential for all participants in collecting networks, allowing them to maintain relationships across distance and often producing valuable information and specimens. In keeping with the shared emphasis on correspondence and visiting, nurserymen and plant hunters alike strove to present themselves as conforming to the cosmopolitan principles that underpinned the Republic of Letters.[17] James Lee, for example, sanctimoniously declared to Joseph Banks and several botanical amateurs that, 'Tho I live by Plants I love to Communicate, for the good of science'.[18] But Lee's apparently impartial cosmopolitan behaviour directly supported his business interests: he relied on the network of collectors to obtain new specimens to sell.

Were Lee and his contemporaries seen as self-interested profiteers, or as impartial scientific practitioners? Most scholars of botany were aware of the duality within nurseries' interests in plants but did not necessarily perceive this to be problematic, and in some cases felt that it could be of benefit to science. In 1780, Chrétien-Guillaume de Lamoignon de Malesherbes, the French minister of state and member of the recently founded Société Royale d'Agriculture, wrote a memo about obtaining rare and valuable plants from America.[19] It began:

Messrs Bartram, botanists . . . of Philadelphia could conduct a very advantageous commerce of seeds of American trees with French merchants, as their father Mr Bartram used to do with the merchants of London.[20]

Malesherbes' note referred to an earlier arrangement between the Philadelphian plant hunter John Bartram and the British amateur of botany Peter Collinson. From 1736 until his death in 1768, Collinson

---

[16] Parsons and Murphy, 'Ecosystems under sail', pp. 505, 512–521.

[17] Moreau, 'Le voyageur français', pp. 148–149, 154–155. On the shifting nature of patriotism and national identity in eighteenth-century France and Britain, start with: Bell, *Cult of the Nation*; Colley, *Britons*.

[18] BL, Add. Ms. 18565, Kaye Notebooks, vol. XVI, f. 83r. Kaye did not date this conversation, but it probably took place around 1790.

[19] Rappaport, 'Malesherbes'. Malesherbes was also member of the Académie Française and honorary member of the Académie des Inscriptions et Belles-Lettres and Académie Royale des Sciences.

[20] AN, 399/AP/98, Chartrier de Malesherbes, Memo [1780?].

and a coterie of wealthy British amateurs had each contributed a subscription to support Bartram's collecting activities in northeastern America. In return, Bartram sent a box of specimens to each subscriber every
year.[21]

Malesherbes hoped to replicate this form of regular exchange by
proposing an agreement between John Bartram's son (also called John,
1743–1812) and Philippe-Victoire Lévêque de Vilmorin. Significantly,
Malesherbes envisaged that scientific exchanges might take place within
the marketplace, rather than under the close control of the state.[22] However, he envisaged a different form of exchange than that established by
Collinson. Malesherbes explained that it would be better for Bartram to
send his plants to a commercial nurseryman such as Vilmorin, rather
than directly to an amateur ('*curieux*') botanist, as Collinson had done:

Commerce will serve them [the Bartrams] much more advantageously than
sending parcels to the *curieux*, because each *curieux* has nothing but the desire
that their own well-being is satisfied.[23]

Worried that the range of plants brought to France might be determined
solely by the fickle desires of modish amateurs, Malesherbes considered
the benefits that traders could bring to science. He noted, first of all,
that merchants who were experienced participants in correspondence
networks would create sustainable relationships, building connections
with a range of reliable overseas suppliers. In addition, they would aim
to satisfy the needs and desires of *all* scholars, not just the whims of a
wealthy minority of *anthophiles*.[24]

Joseph Banks, like Malesherbes, sought connections with the nurseries. He monitored the contents of nurseries in and around London,

---

[21] Bartram sent his first parcel to Collinson in 1734; his friend arranged the subscrip-
tions for Bartram's boxes following complications about remuneration. Bartram and his
sons then continued to send seeds and plants to European subscribers until well into
the nineteenth century. For more on Collinson and Bartram, start with: Bartram and
Marshall, *Memorials*; Cutting, *John and William Bartram*; Grieve, *Transatlantic Gardening Friendship*, pp. 17, 23–24; Swem, 'Brothers of the spade'; Wulf, *Brother Gardeners*,
pp. 19, 29–31.

[22] In other words, beyond the immediate control of the 'colonial machine'. On this concept, see: McClellan and Regourd, *Colonial Machine* and Banks, 'Communications and
"imperial overstretch"'.

[23] AN, 399/AP/98, Malesherbes, Memo [1780?]. See also: AN, AJ/15/511, Envois de
graines, plantes, minéraux etc de pays étrangers au Jardin du Roi, Thouin, Extrait du
Catalogue de Bartrams.

[24] Malesherbes' concerns about the self-interested nature of amateurship were echoed
later in the *Feuille du Cultivateur* in 1793 (discussed in Chapter 3). Heaping praise on
the cultivation of ornamental flowers, Citoyen Gouffier contrasted traders positively to
amateurs. The latter, jealously guarding their exclusive collections, preferred to destroy
their own flowers rather than sharing them with others. *Feuille du Cultivateur*, 30 Janvier
1793, p. 39.

making sure to obtain their most interesting plants. The London nurseryman William Malcolm wrote to him in 1791, for example, to report
that his 'Mimosa Obliqua is now in Compleat Bloom – upwards of 50
Blossoms are now perfectly open... If you should wish to see it, I will
send it any morning'.[25] James Lee even asked Joseph Banks to obtain
specimens for him, in return for gifts of his own plants and other information. In 1775, Lee replied to a query from Banks about prices charged
by nurseries, and concluded his letter by asking, 'Can you procure me
some cuttings or plants of <u>Salix rosmarinifolia</u>'.[26] Lee clearly had no
qualms about asking for this, even though his ultimate intention was to
sell the plant.

Like his French counterparts, Banks believed that the commerciali-
sation of botanical specimens would increase the range of plants that
reached Europe: the specimens that plant hunters sought overseas would
be defined no longer by the fanciful demands of selfish collectors. Further, diversifying the number of people in possession of these rare specimens heightened their chances of survival. Commerce could potentially
benefit science greatly. Banks therefore encouraged the nurseries' participation in cosmopolitan scholarly networks.

The nurseries' customers, as we have seen, responded positively to
the traders' self-promotion as members of the scientific community. The
traders accordingly played down their connections with commerce and
created the impression that their ultimate aim was to contribute to science and thus the general good. Cosmopolitanism added another strand
to this construction. By the early nineteenth century, assertions like that
of James Lee (quoted earlier) that science was more important to him
than commerce were widespread in other nurseries' correspondence and
publications. In 1815, Lee's son unwittingly echoed his father's earlier
claim to scholarly generosity by asserting, quite disingenuously, that 'We
have been too liberal in our pursuits which has prevented our accu-
mulating money,... however... we are happy... even without making
Money.'[27]

Nurseryman Conrad Loddiges conformed to the same pattern. His
six-volume *Botanical Cabinet* (1817) contained several hundred coloured
plates of the flowering plants sold by his nursery, accompanied by short

[25] RBG, Banks Collection, 2.27, William Malcolm (Stockwell) to Joseph Banks, 5 January
1791.
[26] Nottingham Trent University, Sir Joseph Banks Archive Project, James Lee (Vineyard
Nursery, Hammersmith) to Joseph Banks, 5 August 1775 [Original in Yale University
Mss]. I am grateful to Dr Neil Chambers for finding this letter for me.
[27] NHM, H.B. Carter Transcript, James Lee (Hammersmith) to Joseph Banks, 30 August
1815.

written descriptions. In each description, however, Loddiges emphasised above all his friendships within the scientific community. In one typical example, he explained that bulbs of the *Pancratium* 'rotatum' had been sent to him by French plant hunter André Michaux. But rather than describing the characteristics of the plant, Loddiges instead described those of its collector:

Bulbs of this very rare plant were first sent us, about the year 1790, by our very generous friend, the elder Michaux, a botanist who lives in the memory of every one who had the happiness of his acquaintance. Nothing, perhaps, ever equalled the zeal with which this inestimable man laboured, by every means, to diffuse the productions of the various countries through which he travelled for the mutual benefit of each. No selfish view could harbour in his liberal mind; no narrow desire of exclusive possession ever contracted his bountiful heart. The remembrance of such a character is truly refreshing.[28]

Similar testimonials recur throughout the *Botanical Cabinet*. Readers gained little information about the plants concerned, but learnt much of Loddiges' successful integration within the cosmopolitan scientific community.

Historians have presented a contrasting picture of the cultural associations between science and national sentiment. Discussing the extent to which scholars of science in the later eighteenth and nineteenth centuries identified with their own nations, Ludmilla Jordanova concluded that 'practitioners... actively built imagined communities for themselves which were based, more or less, on national boundaries'.[29] But for Loddiges, as late as 1817, the case was exactly the opposite: he singled out individual heroes from a community that was strikingly international.

Commercial nurserymen such as Lee and Loddiges wished to show that, in accordance with the ethos of the Republic of Letters, they placed the cosmopolitan principles of friendship, toleration and cooperation before their business concerns. The nurseries' status within botanical networks could have been compromised by the fact that their need for commercial success was still the driving force behind such claims. However, we have seen that their assertions of cosmopolitanism were not completely disingenuous: plant merchants genuinely welcomed nonfiscal exchanges with other 'scholarly' collectors, for these ensured that they could maintain their positions within the scientific exchange network, and thus enhance their potential to be the first to obtain new plants that could be commercialised.

[28] Loddiges and Sons, *Botanical Cabinet* (1817), vol. 1, description of pl. 19, '*Pancratium rotatum*'. I have not been able to identify this plant by a modern botanical designation.
[29] Jordanova, 'Science and nationhood', p. 197.

This apparent contradiction, between the private interests of the plant traders and their public commitment to the wider scientific community, was in fact paralleled by a similar incongruity at a national level. Scholars in both Britain and France knew that the information they garnered, perhaps especially that which arrived from beyond Europe, could serve national interests. But, as good cosmopolitans, they also espoused a commitment to sharing information and specimens with scholars from other countries. By cultivating and publicly promoting their cosmopolitan identity, the nurseries made a virtue out of a necessity, effectively exploiting the cosmopolitan nature of the Republic of Letters for their own profit. The popularity of the *Botanical Cabinet* and the longevity of Loddiges' nursery suggest that consumers found this cosmopolitan gloss attractive.[30] For them, buying Loddiges' books and plants was a means of buying into science: consumers could thus also participate in a little part of the Republic of Letters. This was, in effect, a way of 'consuming' cosmopolitanism.

Botany on the High Seas

The construction of Enlightenment scientific knowledge, and the expansion of eighteenth-century empires, depended heavily on maritime mercantile networks. Blown by trade winds and carried by currents, merchant ships pitched and plunged through the oceans, linking lands, peoples and objects together.[31] Commerce provided the logic for early empire and, consequently, became closely aligned with imperial and scientific aspirations.[32] Botany, indeed, was a particularly salty science, and botanists and plant traders alike depended on a wide range of overseas connections.[33] How did maritime networks enable collectors to transfer plants between disparate parts of the world?

Mercantile connections structured imperial networks, rather than the other way round. Over most of the eighteenth century, the British and French merchant navies experienced rapid expansion. French overseas trade escalated particularly quickly: between the periods 1716–20 and 1784–88, the average values for France's exports and imports

---

[30] According to the British Library Integrated Catalogue, a total of twenty volumes of the *Botanical Cabinet* were published between 1817 and 1833. Conrad Loddiges purchased a nursery in Hackney from John Busch in 1771. His family ran the business until c. 1860. Harvey, *Early Horticultural Catalogues*, p. 13.

[31] Drayton, 'Maritime networks', pp. 80–81; Killingray, 'Introduction', pp. 3–4.

[32] Gascoigne, *The Enlightenment*, ch. 5; Mackay, *In the Wake of Cook*, pp. 12–14, 168 ff.

[33] There is still a lack of scholarship that looks specifically at the role of the sea in imperial history or the history of science. Exceptions include Killingray *et al.*, *Maritime Empires* and Foxhall, *Australian Voyages*.

quintupled.[34] This was mainly due to the rapid growth of France's colonial trade, which increased tenfold over the same period.[35] Political conflicts had little overall impact on the value of foreign trade to France. Although she lost most of her colonial possessions following the Treaty of Paris in 1763, her import and export trade continued to expand rapidly.[36] By 1787, French imports had been reoriented towards Europe. However, France had not lost all her connections overseas. The surviving French colonies remained major sources of imports, especially Saint Domingue (now Haiti), which provided 34.43 per cent of imports in 1787, compared to a combined total of 31.82 per cent from Italy, Germany and Britain. France's export trade continued to grow healthily, and its non-European markets expanded in particular. The 10 per cent fall in the proportion of imports from France's colonial possessions was balanced by a rise of nearly 7 per cent in exports to colonial markets. Thus, although she lost control of certain key locations overseas, France maintained strong mercantile connections across the globe. Even when she lacked official political control overseas, botanists and plant traders could continue to obtain new specimens via the country's mercantile connections.[37]

French maritime trade never equalled that of the British, however. Despite the increase in her overseas trade during the 1770s and 1780s, France was already struggling in the face of British competition.[38] Britain's overseas trade surged upwards throughout the eighteenth century: that of England and Wales alone rose by a factor of approximately 2.5, from £13 million in 1716–20 to £31 million in the 1780s.[39] By early in that decade, the country's two greatest sources of imports were the British West Indies (24 per cent) and the East Indies (16.9 per cent).[40] The most important non-European recipients of British exports were the United States of America (12.6 per cent of the average annual value of exports), the British West Indies (10 per cent) and the East Indies

---

[34] France's larger population meant that the per capita value of foreign trade was always lower than in England. Butel, *L'économie française*, pp. 80–81; Crouzet, *Britain Ascendant*, p. 18; Jones, *The Great Nation*, pp. xxii, 159–170.

[35] Crouzet, *Britain Ascendant*, p. 29.

[36] Briost, *Espaces maritimes*, pp. 46–47; Jones, *The Great Nation*, p. 244.

[37] Butel, *L'économie française*, pp. 83, 88.

[38] Heywood, *Development of the French Economy*, p. 53.

[39] Butel, *L'économie française*, p. 81; Crouzet, *Britain Ascendant*, pp. 17–18; Rediker, *Devil and the Deep Blue Sea*, Appendix B, pp. 301–303; Schumpeter, *English Overseas Trade Statistics*, pp. 10, 17–18.

[40] Percentages of the average annual values of imports, calculated for the period 1781–85, from figures given in Schumpeter, *English Overseas Trade Statistics*, p. 18.

(7 per cent).[41] Colonies and trading posts naturally became the places that were scoured most thoroughly for new plants.

Plant hunters and specimens travelled along the maritime routes established by global trade. Certain parts of the world, such as Saint Helena and the Île de France (now Mauritius), became entrepôts where plant hunters exchanged specimens and news, and obtained fresh stocks of equipment.[42] European trading companies and their governments soon established a global network of botanical gardens between which plants were transferred, and within which fragile specimens could rest before completing their gruelling voyages around the world.[43] In a 1798 letter to William Roxburgh of the Calcutta Botanic Garden, Joseph Banks described how 'Smith has sent over two Cargoes of Plants for England, the first was left by the Capt at St. Helena to be nursd [sic], . . . [In] the 2^d parcel . . . there were scarce any thing alive in the whole . . . some of them however seem as if they will revive in our Kew Hospital'.[44] Botanical gardens were created in such locations due to the constraints imposed by the biology of plant transfer. But these islands were also sites that were strategically important for colonial expansion and control. Colonial botanical gardens, then, effectively served as locations in which imperialism, science and trade were mapped on to one another.[45]

Trade routes, which developed independently from scientific concerns, primarily determined where botanists could go to collect plants in the late eighteenth century. Although many colonies and trading posts were naturally rich in flora, nurseries and botanists began to complain by the later part of the century that they were exhausted as sources of new plants. Writing to Joseph Banks from Antigua in 1779, plant hunter Francis Masson explained that he had collected 'many plants which appear to me new, But there has [sic] been so many Botanists here of late who I presume have been too industrious to leave any thing for me.'[46] Plant collectors were pushed into less familiar parts of the world.

---

[41] Percentages of the average annual values of exports, calculated for the period 1781–85, from figures given in Schumpeter, *English Overseas Trade Statistics*, p. 17.

[42] Drayton, *Nature's Government*, p. 120; McAleer, 'Young slip of botany'; Roberts, 'Le centre de toutes choses'.

[43] Brockway, *Science and Colonial Expansion*, pp. 75–76; Cook, *Matters of Exchange*, pp. 305–330; Drayton, *Nature's Government*, pp. 120–124; Grove, *Green Imperialism*, pp. 177, 195–197; Mackay, *In the Wake of Cook*, p. 176.

[44] BL, Add. Ms. 33980, f. 159, Joseph Banks to William Roxburgh, 09 August 1798.

[45] Bret, 'Les "Indes" en Méditerranée?', pp. 66–69; Grove, *Green Imperialism*, pp. 168–169, 311; Letouzey, *Le Jardin des Plantes*, pp. 174–178.

[46] PSJB, 5.13.16, Francis Masson (Antigua) to Joseph Banks, 25 September 1779.

Although new specimens were objects of commerce as well as of botanical interest, they were rarely carried on specially commissioned ships. Instead, small parcels of seeds or plants were transported as part of private cargos. It is very difficult to obtain firm evidence regarding the scale of this transfer, as they were not normally entered in official records. Some of the officers on Britain's East India Company ships, for example, occasionally took live plants with them, but evidence for this is very slight.[47] In any case, small weight and space allowances meant that Company servants, like other sailors, were unwilling to carry such frail commodities; those minded towards commerce were much more likely to take compact high-value goods (such as diamonds) in order to maximise their returns.[48]

Plant hunters wishing to send parcels home depended on finding intermediaries to care for the specimens during the voyage. This was easier said than done, because consignments were usually very demanding. It was imperative that specimens were packaged and carried according to precise conventions to ensure that they reached their destinations in a condition that was as close as possible to their state at departure.[49] Seeds and dried plants had to be kept safe from hazards such as heat, saltwater and vermin, and even from sailors, who (supposedly) liked to help themselves to the alcohol used to bottle specimens.[50] Live plants were particularly troublesome: they took up space aboard ship, required fresh water and were more susceptible to decay than most seeds. Plant hunters needed to be confident that a sufficiently attentive intermediary would convey their plants. Just like Thomas Blaikie, they therefore integrated themselves into local scholarly collecting and communication networks, whenever these existed.[51]

It was also important that plant hunters ensured that the ships on which they had entrusted their parcels of plants were destined for the correct port. Overland travel within the country of destination presented new dangers and increased the overall length of the journey. Plants

---

[47] In his research on the East India Company privilege trade, for example, Professor Huw Bowen found that some Company servants did transport live plants, but explained that the evidence was 'circumstantial and based upon the occasional snippet in letters.' H. V. Bowen, Personal communication by email, 13 November 2008.

[48] On the privilege trade, see: Bowen, *Business of Empire*, pp. 247–248, 270–271. On diamonds, see: Vanneste, *Global Trade*, pp. 45–50.

[49] Such objects could be described in Latourian terminology as 'immutable mobiles'. For more on immutable mobiles and the relevant historiography, see Chapter 4, p. 139.

[50] Grieve, *Transatlantic Gardening Friendship*, p. 8.

[51] Christopher M. Parsons and Kathleen S. Murphy have discussed these issues at greater length for the Atlantic. See: Parsons and Murphy, 'Ecosystems under sail', *passim*, esp. p. 507.

destined for Paris were usually sent to the ports of Le Havre or Rouen.[52] In 1786, Thomas Jefferson wrote from Paris to John Bartram in Philadelphia, explaining that Le Havre 'would be the best port to send them [plants] to, because they would come from thence by water. But if no opportunity occurs to that port, let them come... [to] Nantes or l'Orient.'[53] The need to consider which port would receive new parcels added to the complexity of the oceanic transfer of plants, because it was not always possible to find ships taking appropriate routes. Bartram replied to say that 'I have made frequent enquiry after vessels Bound to the ports... mentione[d] in your Letter without success[.] I conclude it is not prudent to send a Large quantity as I don't know of any Convenient opportunity at this time.'[54]

The ports that lined the coasts of each country nevertheless received increasing numbers of new plants, stimulating the development of provincial botanical gardens in maritime towns. Notable British gardens included those at Edinburgh, created in 1764, and Liverpool, created in 1800–02.[55] In contrast to Britain, however, the French provincial gardens, such as those created in Brest (1768) and Bordeaux (1780), did not normally keep Paris informed about their burgeoning collections.[56] By the late 1780s, André Thouin had to resort to sending out enquiries from the Jardin du Roi to find out what they had.[57] The development of botanical gardens in France's ports, then, did not necessarily aid specimen transfer within the country.

The most certain method of transferring plant specimens was to use ships specially commissioned for scientific research and plant transfers. James Cook's ship *Endeavour*, and the *Boussole* and *Astrolabe* commanded by Jean-François de Galaup, comte de Lapérouse, carried

---

[52] André Thouin's correspondence and other records contain many fleeting references to the routes taken by the plants. See especially: AN, AJ/15/149, Dépenses du Jardin du Roi; AN, AJ/15/511, Envois de graines, plantes, minéraux etc des pays étrangers au Jardin du Roi.

[53] LC, Jefferson Papers: Coolidge Collection, Reel 1 1705–1786 Ac 6913, Thomas Jefferson (Paris) to John Bartram, 27 January 1786.

[54] LC, Jefferson Papers: Coolidge Collection, Reel 1 1705–1786 Ac 6913, John Bartram (near Philadelphia) to Thomas Jefferson [Paris], 11 December 1786. Francis Masson was also confronted by the same problem, writing to Joseph Banks in 1784 that 'I would have returned to England by this opportunity [i.e. on the ship carrying his letter] had the vessel not been bound to so remote a port, and so small, that I could not bring home the collection of living plants which I have collected here'. PSJB, 5.13.21, Francis Masson (Madeira) to Joseph Banks, 25 March 1784.

[55] Drayton, *Nature's Government*, p. 134.

[56] Brygoo, 'Jardins royaux et princier', pp. 65–78; Livesey, 'Botany', pp. 57–76.

[57] MNHN, Ms. 308 'Projet d'établissement de correspondence entre les colonies françaises et le Jardin du Roi', January 1788.

specialised containers and made space available to permit the collection of live plants.[58] The ships themselves effectively became floating laboratories.[59]

Perhaps inspired by these expeditions, some amateurs and traders also hoped that it might be possible to commission their own vessels. In 1779, the amateur of botany François Barbé-Marbois (1745–1837), who was the French Consul Général in America, proposed that his countrymen should organise a subscription to hire a ship to convey American plants to France.[60] The plants would, he explained, have a much greater chance of arriving in a healthy state because the ship would be 'uniquely charged with this commodity... and conducted by people who will be especially charged to take care of them'.[61] Although undoubtedly an attractive proposal, nothing came of these plans, probably because of the cost: a single Atlantic voyage could amount to anything from £2000 to £10 000.[62] Ships that were specially fitted out to aid plant transfers constituted a miniscule proportion of the total number at sea. Plant traders, consequently, depended primarily on contact with individual travellers in order to obtain new specimens and to ensure their safe passage across the oceans.

Nurseries' Correspondence Networks

How wide-ranging, then, were the nurseries' contacts, and who collected plants for them? Maps 5.1–5.7 sketch out the international correspondence networks of Andrieux and Vilmorin and Kennedy and Lee between 1760 and 1815. They indicate the distribution of correspondents that each nursery enjoyed, but they cannot represent the actual total of plant hunters and suppliers because the nurseries' correspondence records are incomplete.[63] Instead, they outline the scope of an activity that took place on a much larger scale.

---

[58] For examples, see: Lebreton, *Manuel de Botanique* (1787), pl. 8. Some of the containers shown in Lebreton's book were used on the *Boussole* on Lapérouse's voyage around the world. Ferloni, *Lapérouse*, pp. 28–29.

[59] Banks, *Endeavour Journal*, vol. 1, pp. 22–30; Drayton, 'A l'école des français', p. 96; Ferloni, *Lapérouse*, pp. 22–29; Gaziello, *L'expédition de Lapérouse, 1785–1788*. The phrase 'floating laboratory' is from Beaglehole, *Life of Captain James Cook*, p. 293. See also: Sorrenson, 'Ship as a scientific instrument', pp. 221–236.

[60] Barbé-Marbois lived in Philadelphia between 1779 and 1785. Eugene Parker Chase, 'Introduction' in Barbé-Marbois, *Our Revolutionary Forefathers*, pp. 3–6.

[61] AN, 399/AP/98, M. de Marbois (Philadelphia) to [Abbé Nolin], 15 October 1779.

[62] Rediker, *Devil and the Deep Blue Sea*, pp. 73–74.

[63] I have not found a full run of correspondence for *any* nursery.

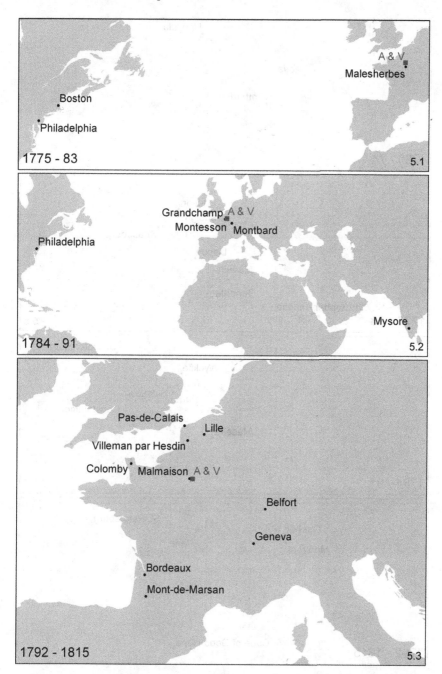

Maps 5.1–5.3 Global distributions of Andrieux and Vilmorin's correspondents, 1775–1815.

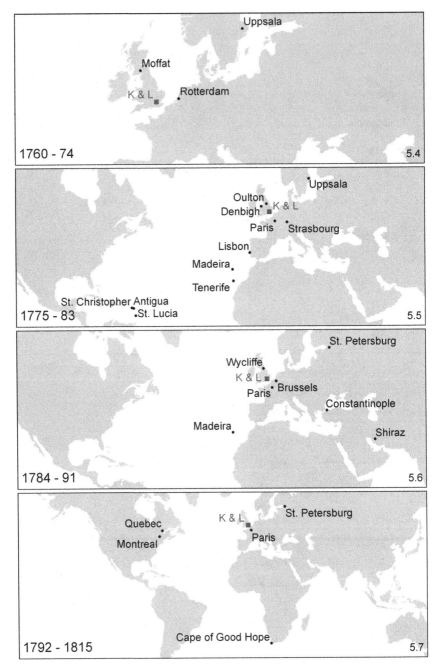

Maps 5.4–5.7 Global distributions of Kennedy and Lee's correspondents, 1760–1815.

Maps 5.1 and 5.2 depict Andrieux and Vilmorin's correspondence network between 1775 and 1791, showing that it stretched from America to India, and possibly beyond. The relatively small number of correspondents shown is almost certainly not representative of the total scale of Andrieux and Vilmorin's global contacts. Their 1778 catalogue, for example, ran to 131 pages and included a high proportion of exotic plants. We also know that they shared many of the same overseas collectors as André Thouin at the Jardin du Roi.[64] Pierre-Marie-Auguste Broussonet (1761–1807), for example, travelled in France and Britain and sent specimens to both the Jardin du Roi and Andrieux and Vilmorin. He also collected for Joseph Banks. André Michaux travelled to Britain, the Middle East and North America, and distributed his specimens likewise.[65]

As well as importing plants and seeds from across the world, Andrieux and Vilmorin redistributed their produce globally via voyages of exploration. The nursery supplied the seeds taken on Lapérouse's circumnavigation of 1785–88 and on the voyage of Aristide Aubert DuPetit-Thouars (1760–1798), who set off in 1792 in search of Lapérouse. The seeds were sent as part of a collaborative experiment on plant transfers, organised by the Jardin du Roi, which sought to understand whether and how plants degenerated once they left Europe.[66] In addition to sharing collectors, then, Andrieux and Vilmorin also participated in experimental research alongside the Jardin botanists. In this respect, Andrieux and Vilmorin could be said to have commanded a scaled-down version of the network that supported the most prestigious botanical garden in France.

Maps 5.4–5.6 show the distribution of correspondents of Kennedy and Lee's Vineyard Nursery, which was possibly more wide-ranging than that of their Parisian counterparts. Kennedy and Lee profited in particular from their close relationship with Joseph Banks. Indeed, the latter often engaged collectors on the strength of James Lee's recommendations. It was Lee, for example, who proposed that Sydney Parkinson should accompany Banks as botanical artist on the *Endeavour*, and in

---

[64] For the geographical diversity of André Thouin's correspondence network, see: Spary, *Utopia's Garden*, pp. 74–75.

[65] The correspondence between Broussonet, Michaux and Joseph Banks is partly summarised in Banks, *Banks Letters*. On Broussonet, see: Caillé, *Un savant montpelliérain* and AN, 399 AP 99, Chartrier de Malesherbes; MNHN, Ms. 314, État de la correspondance de A. Thouin. On Michaux, see references cited elsewhere in this chapter and also: AN, O/1/2113A, Michaux à l'Amérique Septentrionale.

[66] Lapérouse records planting some of his seeds on Easter Island. See: La Pérouse, *Journal*, vol. 1, p. 61. MNHN, Ms. 1945, Expédition d'Aristide Aubert Dupetit-Thouars pour la Recherche de La Pérouse, 1792, ff. 11bis, 13. Such experiments were relatively common among members of the botanico-horticultural network in Paris. See also: BNF, NAF 2578, f. 26, Abbé Nolin to M. le consul de S… [illeg.], September 1781.

1775 he helped to engineer the appointment of gardener David Nelson (d. 1789) on Cook's third voyage.[67] Lee also trained collectors such as George Caley (1770–1829).[68] Many of Banks' correspondents consequently asked to be remembered to 'Mr. Lee' and enclosed gifts for the nursery.[69] The maps show that the people who sent specimens to Kennedy and Lee were located not only across Europe (for example, in Paris, Rotterdam and Strasbourg) but also across the world (from St Lucia to Shiraz to St Petersburg).

Botanical collectors gathered plants from, and communicated with, several different nations at once. André Michaux primarily served French patrons, but we saw earlier that he also sent plants to Conrad Loddiges. Joseph Banks and James Lee, too, received specimens from Michaux: in a letter accompanying a parcel of seeds sent to Banks in 1784, Michaux requested that, 'if you share seeds amongst Cultivators, I recommend to you Mr Lee of Hammersmith ... he has communicated several Plants to me and I wish to show my gratitude to him'.[70] He also reported to French correspondents observations he had made in British nurseries, particularly with regards to cultivation. Writing of the *Ralencia latifolia*, for example, he explained that, 'I have seen the young trees in England in the nursery of M. Gordon[,] who explained to me that he had sown [seeds] fifteen years ago and that they had grown no more than four inches in height, as a consequence they should be sent as [live] plants.'[71] Michaux maintained, and benefitted from, his British connections, even though he was officially working for France.

Horticulture and botany were, consequently, closely aligned in the minds of the plant hunters. Those who had been trained in nurseries understood which sorts of specimens would be of interest to traders. André Michaux, Francis Masson and Thomas Blaikie all described plants' aesthetic attributes in addition to explaining how specimens might be put to use.[72] These collectors possessed an astute

[67] BL, Add. Ms. 33977, James Lee (Vineyard Nursery, Hammersmith) to [Joseph Banks], 26 April 1775. H. B. Carter wrongly dates this as having taken place in 1776. Carter, *Sir Joseph Banks*, p. 172.

[68] NHM, DTC, vol. 10, ff. 157–158, Joseph Banks to George Caley, 7 January 1798; NHM, DTC, vol. 11, ff. 15–16, Joseph Banks to George Caley, 16 July 1798.

[69] For examples, see: BL, Add. Ms. 33977, ff. 20–21, James Manson (Rotterdam) to Joseph Banks, 08 March 1771; BL, Add. Ms. 33977, f. 154, J. Lloyd to Joseph Banks (Revesby Abbey, Lincolnshire), 30 June 1782.

[70] BL, Add. Ms. 8096, f. 137, André Michaux (Shiraz) to Joseph Banks, 15 March 1784.

[71] MNHN, Ms. 357, IX Michaux, Arbres de l'Amérique Septentrionale, f. 3 [undated, 1780s]. I have not been able to identify this plant by a modern botanical designation.

[72] For example, see: PSJB, 5.13.15, Francis Masson (Madeira) to Joseph Banks, 1 February 1779. Masson observed that one particular plant 'is very elegant in a window where it may be trained upon a little lattice work and will give aboundance [sic] of flowers'.

understanding of what sorts of plants would be suitable for commer-
cialisation, and thus most desired by their mercantile patrons.

That the nurseries shared the same collectors as botanists was sometimes misunderstood by botanical amateurs not directly party to the
exchange agreements. In 1802, George Annesley, Viscount Valentia
(1770–1844), wrote while en route to India to warn Joseph Banks that
he had heard on Saint Helena

> that Messrs Lee & Kennedy were regularly furnished with seeds &c by Colonel
> Gordon, who received them from Mr Masson during the whole time that he was
> at the Cape collecting for his Majesty.
>
> Of the truth of this fact I have no doubt...& it accounts for those Nursery-
> men being ever in possession of the most valuable plants sent home from the
> Cape by Masson.[73]

Annesley failed to realise, of course, that Banks was complicit in such
exchanges.[74]

Banks' collusion underlines, further, the existence of multiple
economies within plant exchange networks.[75] The most prized plants,
usually those that were the rarest, were not immediately accorded an
economic value. The small number of botanists with institutional ties,
such as André Thouin and (in a looser sense) Joseph Banks, normally
obtained such rare plants either as gifts or by commissioning collectors.
The elite nurseries used the same means. Both recipients would share
these exceptionally curious specimens with each other. From a horticultural perspective, it was important to distribute specimens as widely as
possible among trusted recipients, to increase their chances of survival.
At this initial stage, then, the price tag subsequently accorded to the
plants mattered little. Trustworthiness and horticultural skill were more
important than other kinds of status in determining who might receive
the precious consignments.

By contrast, aristocratic amateur George Annesley represents a divergent interest group involved in the global collection of plants. As a
curious (and very wealthy) collector, Annesley might certainly receive

---

[73] NHM, DTC, vol. 13, ff. 324–328, Letter written on behalf of George Annesley, Viscount Valentia ('At Sea, Lat. 24, Long. 93') to Sir Joseph Banks, 13 December 1802.
[74] Richard Drayton has claimed that Joseph Banks 'was furious when he heard that the
commercial nursery of Kennedy and Lee had received seeds collected specially for
Kew'. However, Drayton's assumptions about Banks' response were based on Annesley's letter *to* Banks. Annesley (and Drayton) wrongly anticipated that Banks would
share his own horror. Cf. Drayton, *Nature's Government*, p. 46.
[75] On the attribution of value to naturalia once collected, see: Pomian, *Collectionneurs*, esp.
pp. 303–310.

specimens as gifts from other amateurs, might purchase them from merchants or commission his own plant hunters. His scholarly credentials were impressive – he was a Fellow of both the Royal Society and the Linnaean Society – but he gained these positions largely as a result of his social status and the connections and power that this provided, rather than through specific scientific contributions. Annesley was thus not party to the gifting economy developing among plant traders and institutional botanists. His potential entry into that particular circle probably depended more on the skills of his gardener than on his own scholarly expertise.[76]

Maps 5.1–5.7 represent the international connections that each nursery possessed. The botanical correspondence indicates that duplicata were being packaged up and distributed to multiple natural historical institutions and collectors. The nurseries' contacts were, consequently, just as diverse as those of members of metropolitan botanical institutions, such as André Thouin at the Jardin du Roi. Further, the nurseries were generally keen to promote publicly their international connections, over and above other national or political allegiances. But did they continue to do so during times of war?

War at Sea

Scientific practice was heavily influenced by the cosmopolitan ideal of unfettered scientific exchange, but claims to the effect that 'the sciences were never at war' oversimplify the ways in which this ideal was put into practice.[77] Conflict disrupted collecting networks, which were reshaped as political enmities became more pronounced. The two major wars that took place between 1775 and 1815, the American War of Independence (1775–1783) and the French Revolutionary and Napoleonic Wars (1792–1815), pitted Britain and France directly against each other. Each was fought at sea and in colonial settings; each harboured potentially disastrous consequences for the supply of plants, as well as for the survival of the cosmopolitan culture of late eighteenth- and early nineteenth-century science.

---

[76] My analysis of Annesley's position here is influenced by Bourdieu's theory of capital, which explains how individuals can gain positions of social authority through substituting one form of capital (e.g. cultural) with others (e.g. economic and symbolic). Bourdieu, *Logic of Practice*, pp. 108–133. For an analogous example, see: White, 'Purchase of knowledge'.

[77] For claims that the sciences were never at war, start with: Crosland, 'Anglo-Continental scientific relations', pp. 13–22; de Beer, *The Sciences Were Never at War*.

In general, periods of warfare greatly disrupted merchant shipping, causing disorder to the global transfer of plants.[78] The mobilisation of the navy led to intense competition for manpower, as sailors enlisted or were press-ganged into naval service. The number of mercantile ships that sailed was reduced dramatically; those that did ran a heightened risk of being attacked by privateers. As a consequence, the few ships that did set sail were heavily laden. Naval and military manoeuvres also directly impacted upon shipping routes: key ports were blockaded and strategic locations fell into enemy hands.[79] In particular, the British navy seriously disrupted France's Atlantic trade by successfully blockading the country's coastline.[80]

Disruption to mercantile networks was widespread during times of war, but there were of course winners as well as losers. British privateers, and the British economy overall, profited from the 'prize goods' (loot) taken from foreign ships. In 1779, 1781 and 1794, the total values of prize goods imported to Britain were greater than £1 million; in 1794, they amounted to approximately one-third of the total value of imports.[81] Although some merchants lost a great deal of money and resources during wartime, others made massive fortunes from successful war contracts. In the long term, the triumphant power (which was usually Britain) benefited from taking possession of key trading and military outposts, which opened up new maritime routes, new markets and new locations to plunder for natural resources. A 'trade boom' commonly followed periods of international warfare: after the American War, both Britain and France saw significant growth in their overseas commerce.[82] Wars also tended to stimulate the development of technologies that improved the collection and communication of information, such as surveying and mapping, or instruments to measure longitude, which were also of benefit to scientific travellers.[83]

What happened to the nurseries' correspondence networks during periods of wartime? Maps 5.5 and 5.6 depict the shifting geography of Kennedy and Lee's correspondents and suppliers during and after the American War of Independence, and Map 5.7 indicates the effect

---

[78] Schumpeter, *English Overseas Trade Statistics*, p. 7.

[79] Rediker, *Devil and the Deep Blue Sea*, pp. 32, 34–35.

[80] Crouzet, *Britain Ascendant*, pp. 295–317; Heywood, *Development of the French Economy*, pp. 51–52. France also blockaded Britain, but this had a lesser economic impact.

[81] Schumpeter, *English Overseas Trade Statistics*, p. 8. It is unlikely that plant specimens were taken by privateers as prize goods: they were too fragile and required specialised care during voyages. Most sailors were unlikely to appreciate their value.

[82] Butel, *L'économie française*, p. 82; Rediker, *Devil and the Deep Blue Sea*, pp. 35, 303; Schumpeter, *English Overseas Trade Statistics*, p. 11.

[83] Headrick, *When Information Came of Age*, pp. 10–11.

that the French Revolutionary and Napoleonic Wars had on the distribution of their correspondents. The nursery initially maintained a surprising number of international connections despite the American conflict: the range of locations occupied by their contacts even seems to
have increased between 1775 and 1783, reaching as far as the Canaries
and the Caribbean. However, the proportion of the world covered by
their collectors reduced dramatically during the French Revolutionary
and Napoleonic Wars. Kennedy and Lee's correspondents in Map 5.7
are restricted to four locations: Paris, Saint Petersburg, Quebec and the
Cape of Good Hope. Yet, the nationalities of their collectors remained
relatively diverse, and included several French correspondents. Even
though the geographical scope of their activities appears to have been
circumscribed, botanists and traders on both sides apparently upheld
the cosmopolitan principle of political neutrality.

The effects of the wars on Andrieux and Vilmorin were even more
marked. The paucity of archival material makes it hard to be definitive, but it appears that, while Kennedy and Lee's global network survived both periods of war, that of Andrieux and Vilmorin was drastically
reoriented by the Revolution. In Map 5.3, we see that their correspondents dwelt almost entirely within the contours of Napoleon's European
empire.

The impact of war upon mercantile networks was not wholly negative: political realignments could create opportunities for new connections and alliances between nurseries and collectors. The American War
opened up the possibility for new links between France and America, for
example. The Bartram family sought new European customers following
the Declaration of Independence in 1775. Benjamin Franklin, resident
in Paris from 1776 to 1785, had suggested as early as 1777 that Bartram
should 'send the same number of boxes here that you used to send to
England'. He even promised 'to take care of the sale and returns'.[84] This
could only have been a temporary arrangement: Bartram's later agreement with Vilmorin, discussed earlier, was developed shortly afterwards,
in 1780.

Indeed, it was not unusual in the 1780s for French and American amateurs of botany to make connections via diplomatic channels.
Thomas Jefferson became closely acquainted with Malesherbes and
André Thouin during his residence in Paris as the American minister
to France (1784–89). They quickly established a common interest in

---

[84] Benjamin Franklin (Paris) to John Bartram [Philadelphia], 27 May 1777, reproduced
in Bartram and Marshall, *Memorials*, p. 406; Grieve, *Transatlantic Gardening Friendship*,
p. 25.

botany, and worked together to facilitate transatlantic plant exchanges. In January 1786, Jefferson sent a long list to John Bartram, explaining that these were 'plants and seeds which I should be very glad to obtain from America for a friend here whom I wish much to oblige'.[85] Three years later, he was arranging plant transfers in the other direction: Jefferson wrote to Malesherbes to request 'some seeds of the dry rice of Cochin-China'. He hoped that it would be possible to cultivate the rice in one of the Carolinas, replacing a variety 'which requiring the whole country to be laid under water during a certain season of the year, sweeps off numbers of the inhabitants annually with pestilential fevers.'[86] Thanks to the dual layering of scientific and diplomatic networks, many of the plants that travelled between France and America were intended to improve the wellbeing and prosperity of each nation.

Political alliances, even those created for belligerent reasons, could assist collaboration between scholars. Again, many of these scholarly associations were forged through diplomatic connections. In 1788, ambassadors of Tipu Sultan of Mysore visited France seeking military support against Britain.[87] Louis XVI gave them a treasure-trove of diplomatic gifts, ranging from Sèvres porcelain to Vilmorin's seeds.[88] Yet, in spite of the anglophobic backdrop to the exchange, cosmopolitan friendship surfaced again in the arrangements made by the botanists concerning the seeds sent to India. The French and British botanists agreed that, 'in the event of war', the botanical collections should be sent to Joseph Banks at Kew.[89] But such arrangements, while reassuring in principle, could not be guaranteed in practice. The successive Anglo-Mysore wars of the 1790s, which ultimately led to the fall of Seringapatam in 1799, resulted in the dispersal through looting of Tipu Sultan's impressive scientific collections, and the destruction of his gardens under the feet of both armies.[90]

[85] LC, Jefferson Papers: Coolidge Collection, Reel 1 1705–1786, Thomas Jefferson (Paris) to John Bartram, 27 January 1786.

[86] AN, 399/AP/101, Chartrier de Malesherbes, Thomas Jefferson (Paris) to Malesherbes, 11 March 1789.

[87] The ambassadors were not successful in obtaining this support. Jasanoff, *Edge of Empire*, p. 161.

[88] MNHN, Ms. 307, La Luzerne (Versailles) to André Thouin, 8 August 1788; Easterby-Smith, 'On diplomacy and botanical gifts', pp. 191–209.

[89] From summary of letter from Broussonet (Paris) to Joseph Banks, 25 October 1788, in Banks, *Banks Letters*, p. 165. See also: Parsons and Murphy, 'Ecosystems under sail', p. 532, who describe methods of counter-directing packages whereby, should a parcel be captured, a 'second address . . . instructed the enemy's sailors to send the specimens to one of their countrymen'. The latter was then supposed to forward the consignment to the other country when possible. The system was not, however, completely reliable.

[90] Jasanoff, *Edge of Empire*, pp. 171–173; Letouzey, *Le Jardin des Plantes*, pp. 156–172. Neither of these sources actually mentions what happened to the gardens at Seringapatam,

Plant Transfers and the American War

The ways in which war affected plant collecting are apparent from the correspondence that exists between plant hunters and their botanist patrons. These letters indirectly indicate the effects of war on nurseries because, as we have seen, plant traders and botanists shared the same collectors. Plant hunter Francis Masson corresponded with Joseph Banks throughout the American War of Independence and for the first twelve years of the French Revolutionary and Napoleonic Wars. During his travels, which were funded by Kew Gardens and carried out under Banks' direction, Masson also collected plants for Kennedy and Lee, sending parcels to his patron with the instruction that the contents were 'to be divided between Mr Aiton [for Kew] & Mr Lee'.[91] Masson's correspondence thus outlines the implications of war for nurseries as well as botanists, and highlights the practical problems that conflict posed for the safe transfer of plants across the globe. Periods of war, however, also invoked Masson's patriotic commitment to his country, throwing into question his compliance with Banks' cosmopolitan ethos and jeopardising his potential access to plants.

During the American War, Masson worried that the parcels he sent to Britain would be lost because privateers might seize the ships carrying them. When possible, he took the precaution of sending plants on ships owned by neutral powers. Even this was not secure, however. In 1778, he expressed concern about 'a parcel of plants ... directed to Mr Aiton', because, 'being in a Portuguese vessel which had English property on board they ran a risk, as it appears vessels of all nations become a prize where that is found'. Madeira and the Canaries were, he explained, currently 'infested with American cruizers [sic], having taken several of the London traders to this place'. He was only marginally more hopeful that his letter and a 'small parcel of plants, containing 25 sorts' would reach Britain, although they were sent with 'a Privateer belonging to Plymouth'.[92] His concerns were intensified a few months

---

or the fate of the two French gardener-botanists who accompanied the plants. However, an account of the 1792 Anglo-Mysore War mentions the destruction of gardens on the island fort and the desertion of 'fifty-seven of the foreigners in Tippoo's service' in February 1792, some of whom were 'the artificers sent to Tippoo from France, when his ambassadors returned in 1789'. Dirom, *Narrative of the Campaign in India* (1793), pp. 183 (desertion), 186–188, 211 (garden).

[91] PSJB, 13.21, Francis Masson (Madeira) to Joseph Banks, 25 March 1784.

[92] PSJB, 5.13.09, Francis Masson (Puerto de la Orotava, Tenerife) to Joseph Banks, 20 February 1778.

later because France entered into the conflict.[93] By 1780, 'the circumstances of the times' meant that Masson was unable to send any parcels at all, and he was forced to leave 'a number of Plants' with 'a worthy Gentleman . . . who will transmit them by the first opportunity'.[94]

If war disrupted the means of transporting plants across the world, it also placed individual collectors at risk. In May 1778, Masson asked Banks if he could leave Tenerife, and discussed where he should go next. The current conflict meant that certain parts of the world were, in Masson's opinion, out of the question. He explained that, 'We have the disagreeable news that hostilities are commenced between G. Britain and France, which will render any voyage through the W. Indies a little disagreeable.' Masson favoured North Africa because 'I am informed by several Merchants who have travelled in Morocco, that it could be done with great safety'.[95] Travelling there would not only have ensured Masson's personal wellbeing, but would also have meant that he could have sent plants back to Britain more securely because a greater range of routes was available.[96]

However, Masson was not given permission to collect in Africa. Despite his concerns, he was sent to the Caribbean in the autumn of 1778.[97] Banks' decision suggests the strength of his confidence in the political neutrality of science.[98] Yet Masson clearly did not share his patron's impartiality: in September the following year, he asked if he could travel to Barbados, where 'I should be glad to act in a double capacity viz both as a Soldier + Botanist'.[99] Such a combination of roles would have horrified Banks, and Masson's request was not granted. Banks held that it was important that botanical collectors were not connected in any way to political activity. Indeed, botanists were continually at risk of being suspected of espionage, and Banks himself had personal experience of the problems engendered by this association. During his voyage on the *Endeavour* from 1768 to 1771, he had been refused entry to Rio de Janeiro because officials supposed that he intended to spy on them. Banks tried to assure the Portuguese viceroy of his impartiality by professing his cosmopolitanism. 'I am a Gentleman', he wrote,

[93] PSJB, 5.13.11, Francis Masson (Tenerife) to Joseph Banks, 4 May 1778.
[94] PSJB, 5.13.17, Francis Masson (St Christopher) to Joseph Banks, 15 January 1780.
[95] PSJB, 5.13.11, Francis Masson (Tenerife) to Joseph Banks, 4 May 1778.
[96] It should be noted, however, that war even created dangers for plant hunters travelling in the English Channel. Thomas Blaikie was mistakenly imprisoned in the Channel Islands because local inhabitants believed that he was an American spy. Blaikie, *Diary*, 8–16 June 1777, pp. 117–121.
[97] PSJB, 5.13.13, Francis Masson (Madeira) to Joseph Banks, 23 November 1778.
[98] Mackay, *In the Wake of Cook*, pp. 183–186.
[99] PSJB, 5.13.16, Francis Masson (Antigua) to Joseph Banks, 25 September 1779.

'and one of fortune sufficient to have at my own expence [sic] fitted out that part of this Expedition ... which is intended to examine the Natural history of all the Countries where we shall touch.' He reminded the viceroy that, in keeping with the broad-mindedness associated with his socially elevated status, he sought to benefit everyone, rather than only himself or his country: 'His Britannic Majesty was graciously pleas'd to allow [these researches] ... in consideration of the use which from such researches might accrue to Mankind in general.'[100] But despite such winning claims, Banks was not granted access to the shores of Brazil. Desperately keen to see the country and obtain specimens, he resorted to disguising himself and secretly rowing ashore at night.[101]

Elsewhere, assertions of cosmopolitan disinterestedness were more successful. In 1786, Masson used such claims and connections to secure special permission to collect plants at the Cape of Good Hope. The Governor of the Cape 'treated me in the most friendly hospitable manner, but was at a loss how to act regarding my request; as it was ordered by the [Dutch East India] Company that no stranger hereafter should have liberty to explore the country'. Masson obtained permission to travel further inland thanks to support from other members of the botanical community: he presented a letter to the governor 'from the Dutch Embassador [sic]', and a 'Mr Brand came up from False Bay and exerted his influence.'[102]

Banks' role in overseeing and directing Masson's collecting activities ensured that his emissary dutifully enacted the ideal of scholarly neutrality – as suggested by his success in gaining access to the hinterlands of the Cape. But Masson's request to fight also shows that, contrary to the notion of neutrality, he saw no discordance between collecting plants for science and patriotically defending the interests of his country.

Masson's behaviour exposes a tension at the heart of the cosmopolitan ideal: scholars were encouraged to act patriotically, upholding their national interest, while at the same time serving the international scholarly community – a more abstract entity. Much of the current literature on the cosmopolitan ideal and its practice accepts that it involves balancing competing loyalties. Historians have tended to overestimate the

---

[100] Joseph Banks to the Portuguese Viceroy of Brazil, Don Antonio Rolim de Moura, Conde d'Azambuja (Rio de Janeiro), 17 November 1768, reprinted in Banks, *Letters of Sir Joseph Banks*, pp. 4–6.
[101] Banks, *Endeavour Journal*, p. 190, 21–26 November 1768.
[102] PSJB, 5.13.22, Francis Masson (Cape of Good Hope) to Joseph Banks, 21 January 1786.

significance of this, viewing examples of dual loyalties as exceptional, rather than the norm.[103] But the conflicts and dualities experienced by the individuals who have peopled these pages lay at the heart of the ideal and its articulation. The negotiation of contradictory allegiances was a feature of 'cosmopolitan' cultures, and the cosmopolitan ideal provided Enlightenment scholars with a common identity and purpose.[104]

Plant traders and botanists shared the same aspirations towards international exchange, and generally worked together to achieve this goal. The plant trade, then, reinforced the infrastructure through which specimens and information were exchanged, an infrastructure that was crucial to the wider development of botanical and horticultural knowledge. As Chapter 6 will show, however, the major ideological shifts brought about by the French revolutionaries and then by Napoleon meant that cracks finally began to appear within the cosmopolitan ideal.

---

[103] Edmond Dziembowski, for example, presented the early eighteenth-century French elite's fascination with foreign culture as a 'cosmopolitan veneer', behind which lurked 'a more "national" reaction'. What Dziembowski termed a 'veneer' was much more substantive among Enlightenment scholars. Dziembowski, *Un Nouveau Patriotisme*, p. 165; Nussbaum, 'Kant and Stoic cosmopolitanism', pp. 6, 9.

[104] Withers, *Placing the Enlightenment*, pp. 29–33.

The cosmopolitan ideal was largely upheld during the 1770s and 1780s. During the American War of Independence, plant hunters such as André Michaux fostered connections with botanists and plant merchants located in enemy countries. The French Revolutionary and Napoleonic Wars, however, imposed much more significant practical constraints on scholarly communication: Andrieux and Vilmorin's correspondence network was reduced to the contours of the Napoleonic Empire by the early nineteenth century. Did this contraction reflect a shift in attitudes?

The French Revolutionary and Napoleonic Wars heralded new, overtly nationalist sentiments in France and Britain. David Bell has described how, in France, xenophobic feelings intensified and 'the cosmopolitanism so often associated with eighteenth-century French culture abruptly disappeared from books and periodicals, to be replaced by snarling hostility to France's enemies'.[1] In order to understand how the French Revolution affected cosmopolitan attitudes and international contact, we will first consider the substantive structural and ideological changes that emerged during the 1790s, and how these affected botanists and plant traders within France. We will then return to look at international exchanges during wartime.

Science in Revolution

The fortunes of the botanical scholars and plant merchants discussed in the previous chapters ebbed and flowed as political circumstances shifted during the Revolution and under Napoleon. Most found that things initially remained relatively unchanged: institutions such as the Académie des Sciences and Société Royale d'Agriculture, for example, operated more or less as normal between 1789 and 1792, meeting regularly, publishing *mémoires* and distributing prizes. But the fortunes of these major

---

[1] Bell, *Cult of the Nation*, pp. 82, 99–105, quote p. 82; Colley, *Britons*, pp. 287–289.

institutions shifted dramatically over the 1790s, and this had a heavy impact on French science.[2]

Although the broader changes brought about by the Revolution were widely felt and commented on from 1789 onwards, scholars and amateurs of science were not initially criticised for their scientific activity. The early years of the Revolution were largely concerned with forming and consolidating France's new identity as a nation, renegotiating its relationship with the monarchy and the Catholic Church, and, from spring 1792, combating the threat of foreign invasion. Science did not play a particularly high-profile role in these affairs. From September 1792, however, the government (known from this point as the Convention, and primarily dominated by the uncompromising 'Jacobin' faction) became increasingly concerned with imposing ideological unity on the new nation. Individuals or groups that did not appear to conform to the correct ideological principals were singled out as 'enemies' of revolutionary France. As anxieties about the internal threat posed by these so-called enemies spiralled, the Jacobin-led Convention inaugurated the infamous 'Reign of Terror' (1793–94), during which many of France's internal enemies were imprisoned and even executed by guillotine.

The Jacobin attempts to detoxify France ushered in a deep suspicion of savants. Those who studied pure sciences such as mathematics were accused of pursuing subjects that, as they appeared to lack utilitarian value, were futile lines of enquiry. We saw earlier that scholarship in the Old Regime had been culturally associated with the upper social echelons. The Jacobin critique of scholarship, framed in Rousseauvian terms and fanned by the incendiary writings of journalists such as Jean-Paul Marat (as well as by existing tensions among savants), argued too that indulging in such pastimes had a noxious moral effect on society at large.[3] On 8 August 1793, a resoundingly unsympathetic Convention closed the Académie des Sciences – a move that affected only around fifty savants directly, but that undermined science throughout France and sent shockwaves across Europe. The Enlightenment scientific community had lost one of its principal founding institutions.[4]

The anti-intellectualism that characterised the Terror certainly extended to suspicions about natural history, but naturalists, especially those at the Jardin du Roi, made a strong case for the utilitarian and

---

[2] On the sciences during the French Revolution, start with: Alder, *Measure of All Things*; Chappey, *Naturalistes en Révolution*; Dhombres and Dhombres, *Naissance*; Gillispie, *Science and Polity in France*.

[3] Dhombres and Dhombres, *Naissance*, ch. 1, esp. pp. 30–31, 35–36; Gillispie, *Science and Polity in France*, pp. 189–192.

[4] Dhombres and Dhombres, *Naissance*, pp. 12–14, 35.

patriotic value of their work.[5] Nature itself was selectively exploited for political purposes in the Revolution: natural objects were deployed symbolically to support the new regime, for example as liberty trees (which were supplied by the Jardin from Year II (1793–94) onwards) and in the names given to the new calendar.[6] Natural history promised economic benefits to agriculture, industry and medicine; the study of nature, properly undertaken, might even contribute to the 'regeneration' of the French people. Taken together, these political articulations suggested that the natural world could offer a basis upon which the new nation might be formed. The Jardin botanists consequently escaped from the Jacobins' otherwise fierce assault on scholars. Paris' botanic garden was the first national scholarly institution to be founded in revolutionary France, on 10 June 1793. The professors employed at the newly named Muséum d'Histoire Naturelle were tasked in particular with improving agriculture, arts and trades – utilitarian aims that fit closely with the Jacobin agenda.[7]

The ending of Terror in 1794 and the dissolution of the Convention the following year heralded a new start for revolutionary France. First under the Directory and then (from 1799) under Napoleon, ministers endeavoured to stabilise the country by shunning anything that smacked of Jacobin excess. Savants of all stripes slowly regained social standing; natural history, in spite of its former associations with Terror, remained aligned with the interests and identity of the French nation. Indeed, the study of nature offered an entrancing combination of national and personal prestige, as well as promising the wider economic and imperial benefits noted before. It consequently occupied a prominent place (and thus gained further official endorsement) within the newly devised national curriculum.[8] Under Napoleon, scholars' professed commitment to rationality and human progress meant that those who had survived the Terror were ultimately considered model citizens.[9] Napoleon also developed a new ethos of service to the nation, a move that left little space, however, for the older, cosmopolitan Republic of Letters.[10]

---

[5] Lacour, *La République Naturaliste*, pp. 16–19.

[6] Corvol, *Nature en révolution*; Miller, *Natural History of Revolution*; Perovic, *Calendar*, ch. 3, esp. pp. 117–124.

[7] Gillispie, *Science and Polity in France*, p. 175.

[8] Spary, *Utopia's Garden*, pp. 212–216, 221–227 and 'Forging nature', pp. 168–169.

[9] Chappey, *Société des Observateurs*, p. 52.

[10] Daston, 'Nationalism', pp. 102–112. The projects for overseas voyages of exploration exemplify the shifting attitudes towards science during this period particularly clearly. See: Harrison, 'Projections of the revolutionary nation', p. 36; Jangoux, 'L'expédition aux Antilles', p. 43; Starbuck, *Baudin*, pp. 4–6.

Revolution in the Nurseries

Just as they had done during the Old Regime, the upper-end plant traders initially aligned themselves with the activities of institutional botanists. This required some careful negotiation, however. It was important to shed the associations with the elitist scholarly model that, as we have seen, traders such as Vilmorin had formerly cultivated so carefully. Many of the top French nurseries, furthermore, had served a predominantly aristocratic clientèle, and needed to find new custom. Several of Vilmorin's acquaintances, including Christophe Hervy, the head gardener of the Chatreux monastery, and Thomas-François de Grace (1713–1798), the editor of the *Bon Jardinier*, were ruined by the Revolution, and relied subsequently on Vilmorin and others for support.[11] Those that survived, not only Vilmorin, but also Jacques-Louis Descemet (1761–1839), head gardener of the Jardin des Apothicaires, and Jacques-Martin Cels (1740–1806), a highly regarded amateur of botany, managed by diversifying the plants sold, or undertaking duties that served revolutionary ends.

Cels, Vilmorin and Descemet were also no less keen than the Jardin botanists to display their patriotic commitment to the new nation. Engaging explicitly with the revolutionary valorisation of utility, they participated in programmes to enhance the nation's collection of useful plants, especially through experiments in naturalisation of non-native plants, and in the identification of their useful properties. The contributions to the nation made by these plant traders and amateurs fell largely within three areas: as altruistic distributors of agricultural seeds; as authorities on the cultivation of these and other plants; and as the tireless (and selfless) introducers of new species useful to the nation. Accounts of their patriotic activities were publicised via journals including the *Journal de Paris*, the *Mémoires d'Agriculture* and the *Feuille du Cultivateur*. Their names also appeared within official correspondence concerning agriculture and botany. The endurance of certain individuals and their gardens through the revolutionary and Napoleonic years attests to their ability to adapt their public profiles, and the contents of their gardens, to fit new constructions of natural history and its social value.

Amateur botanist Jacques-Martin Cels was the least typical of the three examples. Cels negotiated a tricky position, as proprietor of what by 1794 would be known pejoratively as a 'jardin de luxe'. Having lost his former employment as customs collector on the Barrière Saint Jacques,

---

[11] Silvestre, 'Notice biographique', p. 209.

he improvised a new income by selling rare plants from the extensive collection in his garden, placing emphasis on those that were in some way utilitarian.[12] In Year 2 (1794), he published a carefully worded article in the *Feuille du Cultivateur* discussing the value to the nation of 'luxury gardens' such as his own. After politely applauding the patriotic work done by zealous citizens who had already uprooted some such gardens to make space for crops such as grains and potatoes, Cels questioned the contemporary presumption that pleasure gardens were indeed useless. The mature trees that made up the majority of such gardens, he pointed out, actually performed an essential role in purifying the air and in providing useful products such as resins, essential oils, nuts and animal fodder. Replacing such resources would take years. Further, he emphasised, the land required for arable cultivation was relatively small: it was much better to exploit a small area efficiently than to attempt to grow crops over a large, unmanageable extent. He concluded by offering to supply young Scots pines and spruces at very reasonable prices, trees that would grow 'in poor soils' and that promised high economic returns.[13] Cels thus publicly aligned his own interests (and the contents of his garden) with those of the nation, devising patriotic schemes that would coincidentally bring him a financial income.

The 1790s also offered Vilmorin an exceptional opportunity to enhance his reputation. Like Cels, he presented his work as serving the national interest. Cels and Vilmorin both also capitalised on their earlier reputations as botanical and horticultural experts. Cels had been well regarded among scholarly circles prior to the Revolution, and had enjoyed a close association with the Jardin du Roi, as mentioned in Chapter 2.[14] Vilmorin consolidated the reputation that he had begun to establish during the 1780s as an enlightened amateur scholar. Both became respected within government circles and among the wider public, and thus gained election to key revolutionary committees, as will be discussed later.

[12] Jacques-Martin Cels maintained a private garden beside the Barrière Saint Jacques from at least 1778 until 1791. As discussed in Chapter 2, amateurs were strongly encouraged to visit it prior to the Revolution. Cels moved to Montrouge by 1792. See: de Grace, *Bon Jardinier* (1785), p. 195; MNHN, Ms. 314, État de la Correspondance d'André Thouin; BL, Add. Ms. 8097, ff. 393–394, Cels (Paris) to Joseph Banks (London), 7 March 1791; BL, Add. Ms. 8098, Cels (Plaine de Mont Rouge, Paris) to Joseph Banks (London), 29 March 1792. The contents of Cels' Montrouge garden were described by botanist Étienne Pierre Ventenat and illustrated by Pierre-Joseph Redouté, in *Description des plantes nouvelles* (1799), and were itemised in an inventory made after his decease in 1808: AN, ET/XLIV/0765, Inventaire après décès de Jacques Martin Cels, 27 June 1808. See also: Spary, *Utopia's Garden*, p. 56, n. 23.
[13] *Feuille du Cultivateur*, Tome IV, Duodi 22 Ventôse, An 2, pp. 93–95.
[14] Spary, *Utopia's Garden*, pp. 56–57.

One of the best summaries of Vilmorin's manifold contributions was published as early as autumn 1789, when the Société Royale d'Agriculture awarded him a gold medal,

for having presented various very interesting observations; having made a great number of useful experiments; placed the Society in a position to distribute valuable seeds; gave free seeds to hard-up Cultivators... to sow in their lands devastated by the hailstorm [of 13 July 1788], [and] sacrificed his right to commission for a very considerable quantity of seeds that the Administration had required him to obtain from abroad.[15]

During the years that followed, both published and manuscript references to Vilmorin continued to emphasise his benevolence, his knowledge and his patriotic selflessness. Vilmorin's generosity to farmers whose crops had been destroyed by the 1788 hailstorm was reported in the *Journal de Paris*, which also noted that the Société d'Agriculture had in addition commissioned Vilmorin to obtain more seeds from foreign suppliers.[16] Further, it was Vilmorin, apparently, who alerted the Society in 1791 to the plight of farmers in the village of Boissy-sous-Saint-Yon whose crops had been destroyed by late frosts, prompting the free distribution of haricot beans among these cultivators.[17]

Vilmorin's reputation as an authority also grew. In May and June 1792, he was thanked publicly by André Thouin for supplying manuscript memoirs on the cultivation of pineapples.[18] The chemist and agronomist Antoine-Alexis Cadet de Vaux (1743–1828) also cited Vilmorin as a respected source of dictums on the cultivation of vegetables, repeatedly quoting his maxims about sowing seeds in 1792.[19] Cadet subsequently replicated verbatim long passages from his correspondence with the nurseryman. Writing about how to rescue artichoke crops lost to frost in spring Year 3 (1795), for example, he asserted that 'I will be nothing but his echo; here is his note that I copy...'[20]

Vilmorin's third major contribution was as a supplier of useful seeds, especially those from abroad.[21] Although most of this book has focused on the traders' roles as merchants of ornamental plants, it was largely through the introduction and naturalisation of agricultural specimens

---

[15] *Mémoires d'Agriculture, d'économie rurale et domestique*, Automne 1789, p. x.
[16] *Journal de Paris*, 23 July 1788, pp. 897–899; 27 July 1788, p. 911; 30 July 1788, pp. 922–923.
[17] *Mémoires d'Agriculture*, Été 1791, p. xi, and Automne 1791, pp. 41–42; *Feuille du Cultivateur*, Tome I, Mercredi 20 Juillet 1791, p. 332.
[18] *Feuille du Cultivateur*, Tome II, Samedi 12 Mai 1792, p. 150.
[19] *Feuille du Cultivateur*, Tome II, Samedi 2 Juin 1792, p. 174.
[20] *Feuille du Cultivateur*, Tome V, Duodi 12 Ventôse, An 3, p. 86.
[21] *Feuille du Cultivateur*, Tome III, Septidi 7 Frimaire, An 2, p. 383.

that Vilmorin secured his reputation in the Revolution.[22] Like Cels, he suppressed associations with opulent, beautiful plants in favour of patriotic specimens associated with utility. Thanks to the connections he had developed prior to the Revolution, Vilmorin was able to furnish France with specimens that were rare or otherwise very difficult to obtain. In Prairial Year 3 (May 1795), for example, the minutes from the Comité d'Agriculture et des Arts noted that the Andrieux and Vilmorin nursery was undertaking a commission for the nation, procuring hemp seeds from a Dutch seed merchant.[23] This type of assignment was not unusual: Vilmorin was one of a small number of Parisian nurserymen regularly celebrated for his work introducing and developing methods for cultivating new species of plants, especially fodder and food crops.[24]

The fortunes of the Descemet family offer an almost direct contrast to those of Vilmorin and Cels.[25] Since the 1760s, Jacques-Louis Descemet's mother had run a commercial nursery close to the Jardin des Apothicaires, from which she and her family sold exotic plants and medicinal herbs, and where they had taught public courses in botany. Descemet had competed unsuccessfully against Jean-Baptiste Lamarck for appointment to the Botanical Section of the Académie des Sciences in 1779. Lamarck had been selected even though Descemet had won the majority of votes.[26] During the Revolution, the Descemet family retained its existing specialism in cultivating ornamental plants and also donated seeds to the Société d'Agriculture for free distribution.[27] Indeed, in 1791, Descemet and Vilmorin were lionised by the Société as two of only six people who regularly supplied it with seeds *gratis*, for free distribution among needy farmers.[28] But Descemet otherwise responded to the Revolution in a manner quite distinct from Vilmorin. Although he had

---

[22] Vilmorin was, nevertheless, also celebrated as one of only three plant merchants who introduced new ornamental plants to France. *Feuille du Cultivateur*, Tome III, Mercredi 30 Janvier 1793, p. 39.

[23] Gerbaux and Schmidt, *Procès-Verbaux*, vol. 3, pp. 484–485.

[24] *Mémoires d'Agriculture*, Printemps 1788, pp. 112–118; *Feuille du Cultivateur*, Tome I, Mercredi 27 Octobre 1790, p. 27; Mercredi 9 Février 1791, pp. 145–146; Mercredi 23 Mars 1791, p. 196; Tome II, Samedi 11 Février 1792, p. 46; Samedi 3 Mars 1792, p. 71.

[25] Joyaux, *La Rose*, p. 21.

[26] Their garden was on the rue des Charbonniers. *Journal de Paris*, no. 257, 14 Septembre 1779. I am grateful to Professor Bruno Belhoste for sending this reference. *Avant-Coureur*, vol. I (1760–1766), pp. 197–8 (1761), pp. 373–4 (1766); Joyaux, *La Rose*, p. 21. On the election to the Académie des Sciences, see: Gillispie, *Science and Polity in France at the End of the Old Regime*, p. 162.

[27] *Feuille du Cultivateur*, Tome 1, Mercredi 9 Mars 1791, p. 180.

[28] *Feuille du Cultivateur*, Tome I, Mercredi 9 Mars 1791, p. 180; *Mémoires d'Agriculture*, Automne 1791, p. 42. The six donors were: Thouin, Parmentier, Cretté, Vilmorin, Descemet and Costel.

formerly occupied a relatively public role, advertising botany courses taught from his garden on the rue des Charbonniers, he shrunk into the shadows during the 1790s. His name seldom appeared in the *Feuille du Cultivateur*, and when it did, it was in the capacity of a supplier of seeds and plants, and not in relation to his botanical expertise. Descemet resigned from the Jardin des Apothicaires in April 1793, and the family disappeared completely from the public eye during the Terror, moving out of Paris to Saint Denis (then temporarily renamed the Franciade). After over a year's silence, they finally issued a public advertisement of their wares, sold by their new suburban nursery, in Pluviôse Year 3 (January 1795).[29]

Negotiating the right level of public exposure was an issue throughout the 1790s for Descemet, Vilmorin and Cels. It seems that each sought social respect by avoiding being too ostentatious, but had to balance this against an economic need to promote their wares. It was important to cultivate an image of modesty and self-effacement, but nevertheless to ensure that one's name and products were well known. Thus, Vilmorin's articles were rarely published under his own name, and he was seldom publicly credited as the source of the letters he had apparently forwarded to the *Feuille du Cultivateur*.[30] He did, however, ensure that other established botanical authorities mentioned his name in print.[31] By contrast, Jacques-Martin Cels contributed under his own name to the *Feuille*, and his articles were invariably followed with a short note informing readers that the plants discussed could be purchased from his garden. He published much less frequently than Vilmorin, however.[32] Jacques-Louis Descemet avoided printing anything in the *Feuille* under his own name, with the single exception of a 1791 article about the beautiful Cytise des Alpes (*Laburnum alpinum* (Mill.) J. Presl).[33] All three ostensibly remained out of the public eye during the 1790s, but the activities of Vilmorin and Cels, as horticultural and botanical mentors, ensured

---

[29] *Feuille du Cultivateur*, Tome V, Septidi 17 Pluviôse, An 3, p. 60.

[30] *Feuille du Cultivateur*, Tome VI, Septidi 17 Fructidor, An 4, pp. 293–295. Vimorin's reticence was also noted by Silvestre. Apparently, the nurseryman had written a great number of essays, reports and instructions, but these were largely unpublished. Consequently, 'sa manière de favoriser les progrès de la culture est moins brillante que la publication d'ouvrages estimables, mais elle n'est pas moins efficace'. Silvestre, 'Notice biographique', p. 200.

[31] *Feuille du Cultivateur*, Tome V, Duodi 12 Ventôse, An 3, p. 52. Vilmorin did, however, join with Cels, Thouin and others as an acknowledged contributor to the less frequently published *Annuaire du Cultivateur*.

[32] *Feuille du Cultivateur*, Tome II, Mercredi 7 Mars 1792, pp. 74–75; Tome V, Duodi 22 Nivôse, An 3, pp. 25–26; Tome V, Septidi 7 Ventôse, An 3, p. 80

[33] *Feuille du Cultivateur*, Tome 1, Samedi 5 Février 1791, pp. 141–142.

that their names appeared regularly but incidentally, via other individuals' publications. Readers of the *Mémoires d'Agriculture* and the *Feuille du Cultivateur* were thus consistently directed to both gardens.

It would be putting it too strongly to claim that Vilmorin actually eclipsed his peers, but it is clear that he occupied a privileged position in the eyes of the authorities.[34] In 1793, the *Feuille du Cultvateur* printed an extract from the minutes of a recent public meeting of the Conseil du Département de Vosges. The Conseil had set aside the princely sum of 1000 livres to allow local farmers to buy seeds from plant traders. The minutes recommended only Vilmorin of Paris by name; the phrase 'or all others' was tacked on as a tardy acknowledgement of the existence of other merchants.[35] Other articles underscored Vilmorin's trustworthiness as a supplier. Later that same year, for example, a correspondent writing about turnips explained that he sourced his genuine 'English' seeds from Vilmorin; he emphasised their authenticity by noting 'the scrupulous exactitude with which Citoyen Vilmorin treats the people who supply seeds to his store'.[36] In these examples, Vilmorin was always a supplier or middleman, whose service to the nation was in line with his own commercial occupation. A footnote to the Vosges article explained to readers that Vilmorin was a 'Famous Botanist and Seed Merchant',[37] and he was later described as an 'excellent practitioner, ... continuing the commerce of all species of seeds, to which he owes one part of his reputation'.[38]

Vilmorin's growing public profile was paralleled by appointment to official academic and government roles. Again, he outshone Jacques-Louis Descemet, gaining appointments of an equivalent standing to that of Jacques-Martin Cels and other respected agricultural authorities. Vilmorin was elected as a Correspondent of the Société Royale d'Agriculture in spring 1789 (along with George Washington, who was made Associé étranger). Two years later, in autumn 1791, he was made Associé ordinaire.[39] Jacques-Louis Descemet was elected as Correspondent in spring that same year, but did not obtain further distinction, in

---

[34] Nevertheless, circumstances were clearly not easy: Several members of the d'Andrieux family (presumably not working in the plant trade) emigrated in Year IV (1795–96) to join relatives already settled in Sainte-Anne de Guadaloupe. See: de Vilmorin, 'La famille ANDRIEUX'.

[35] This is additionally significant given that Paris is around 200 miles from the Vosges. *Feuille du Cultivateur*, Tome III, Samedi 26 Janvier 1793, p. 35.

[36] *Feuille du Cultivateur*, Tome II, Septidi 7 de Frimaire, An 2, p. 383.

[37] *Feuille du Cultivateur*, Tome III, Samedi 26 Janvier 1793, p. 35, fn 1.

[38] *Feuille du Cultivateur*, Tome VI, Septidi 17 Fructidor, An 4, p. 293, fn. 1.

[39] *Mémoires d'Agriculture*, Printemps 1789, p. v; Automne 1791, p. viii.

this or other committees.[40] In 1793, Vilmorin was appointed head of a subdivision within the new Commission d'Agriculture et des Arts, whose scope included rural economy and manufacturing. Directing the third section within Division Two, 'Agence végétale', Vilmorin oversaw matters relating to fodder, vegetables, fruit trees, vines and the production of wine and spirits.[41] He also supervised the management of estates seized by the nation, and in 1794 joined with colleagues in an attempt to save the remarkable *pépinière* of the Chartreux monastery from destruction.[42] Jacques-Martin Cels was made head of the fourth section, 'Amélioration des bois et forêts' (which was also responsible for conservation of and experimentation on rare and exotic plants). To indicate the level of status required for these positions, it is worth noting that the respected agricultural reformer Antoine-Augustin Parmentier (1737–1813) led the fifth section, 'Plantes économiques'.[43]

Nurseries and Nationalised Science

So, Citoyen Vilmorin gained unrivalled opportunities during the Revolution to obtain ever-higher positions within government committees, strengthening the alliance between state agendas and his private commercial activities. What were the consequences of this patriotic activity for Vilmorin's international connections? Maps 5.1–5.3 suggest that his overseas contacts were realigned during the 1790s and that, by the early nineteenth century, they were restricted to Napoleon's European Empire. However, between 1793 and 1829, the only surviving records of Vilmorin's foreign correspondence are with botanists in Switzerland (1806 and 1829). No letters or memoranda remain in the Andrieux and Vilmorin archive to confirm that their correspondence really became confined to the Empire and neutral European countries.

Other evidence suggests that Vilmorin's relations with Britain did not in fact fade away during the early nineteenth century. This was despite the rise of patriotic fervour and mutual hostility between France and Britain. In his *Éloge* to the nurseryman of 1805, Augustin François de

[40] *Mémoires d'Agriculture*, Printemps 1791, p. vii.
[41] *Feuille du Cultivateur*, Septidi 27 Brumaire, An 3, p. 402.
[42] *Procès-Verbaux du Comité d'Agriculture et des Arts*, Vol. 3, 27 Pluviôse An III, p. 393, Art. 6. The Commission ultimately managed to save the trees, but not the plants, from the Chartreux property, whose garden was eventually made part of the Jardin du Luxembourg. Joyaux, *La Rose*, p. 74; Silvestre, 'Notice biographique', p. 201.
[43] *Feuille du Cultivateur*, Tome IV, Septidi 27 Brumaire, An 3, p. 402. The membership and activities of the various revolutionary committees seeking to improve French agriculture have been discussed more extensively in: Lacour, *La République Naturaliste*, pp. 333–353.

Silvestre, secretary of the Société d'Agriculture, had no qualms about celebrating Vilmorin's numerous international correspondents. He even placed special emphasis on his many associations with Britain.[44] This was in keeping with the tenor of the articles in the *Feuille du Cultivateur*, which had also recognised and valued the connections maintained by Vilmorin, Cels and others across the Channel.[45] Silvestre's openness with regards to Vilmorin's contact with Britain is particularly noteworthy because of its timing: he presented the *Éloge* to the Société d'Agriculture on 17 November, barely a month after France's crushing defeat by the British at Trafalgar.[46] Silvestre's account suggests that any contraction in Vilmorin's correspondence network was more due to the practical difficulty of communicating beyond the Empire during wartime than to a substantive shift in attitudes. It is likely that Vilmorin maintained contact with foreigners whenever possible, invoking cosmopolitan values when he did so.

Napoleon's reorientation of French science did not destroy cosmopolitan culture in France, but it did alter French interactions with overseas correspondents.[47] Wartime conflicts disrupted communications between savants, and imperial expansion meant that it was easier for practitioners to make connections with people who lived within the Empire. This conformed to Napoleon's vision for imperial science, about which he claimed famously that 'All men of genius, all those who have attained a distinguished rank in the Republic of Letters, are French, in whatever country they happen to be born'.[48] Napoleon's assertion of the Frenchness of science flew in the face of cosmopolitanism, yet was not mirrored by a general shift in attitudes among French scholars. Vilmorin maintained international connections beyond the French Empire, although he avoided broadcasting these too widely. Cosmopolitan sentiments lost their public appeal, but they continued to frame the international exchange of information between scholars.

The Revolutionary and Napoleonic Wars had a similarly mixed influence upon British attitudes towards cosmopolitan science. As President of the Royal Society, and as a central figure within natural history networks, Joseph Banks enjoyed tremendous influence over British scholars. Tracing the evolution of his attitudes, and of those of people

[44] Sarrut, *Biographie*, vol. 4, pp. 551–552; Silvestre, 'Notice biographique', pp. 195, 197.
[45] *Feuille du Cultivateur*, Tome II, Septidi 7 de Frimaire, An 2, p. 383; Lacour, *La République Naturaliste*, p. 250.
[46] Bret, 'Des "Indes" en Méditerranée?', p. 70.
[47] See also: Daston, 'Nationalism', pp. 104–105.
[48] Napoleon Bonaparte to Bartolomo Oriani, 5 Prairial Year V (24 May 1796), quoted in Daston, 'Nationalism', p. 95.

connected to him, offers a clear indication of wider shifts in British scholarly culture. Banks worked hard to maintain good relations with his French counterparts throughout the 1790s.[49] His stance towards the enemy nation gradually hardened as the war dragged on, however. By the early nineteenth century, his confident assertions of cosmopolitanism, and thus those of his acolytes, had became much more muted.

The wars inevitably constrained the circulation of correspondence and objects, and plant hunters in particular struggled under the practical constraints imposed by conflict. In Chapter 5, we met the plant hunter Francis Masson, who, by 1792–94, was stationed at the Cape of Good Hope. 'I am truely [sic] sensible of the direful effects of the present war', Masson wrote to Joseph Banks in June 1794, 'Science will no doubt greatly suffer by it'.[50] By October of that year, he was complaining that 'the war has so interrupted our communication that I have had no accounts either from Soho Square [i.e. from Banks] or Kew for two years and a half'.[51] As during the previous conflicts, letters and specimens were placed at greater risk due to the increase in privateering: Banks annotated on one of Masson's letters from 1793 that a ship carrying a box of 314 sorts of seeds 'was taken by the French', and the collection entirely lost to British science.[52] The reduced number of merchant ships caused further problems for Masson: the ships that passed the Cape were so crowded that it was not possible to send home 'any considerable collection'.[53] He was forced to entrust his parcels to 'passengers on board Foreign ships who could not take care of any thing large'.[54] Most of these ships were destined for Dutch ports, further prolonging the time taken for specimens and seeds to reach Britain, and reducing their chances of survival.

The channels of scientific communication contracted further as relations between France and Britain worsened over the 1790s, although Banks remained curiously resistant to acknowledging this. Despite the ongoing hostilities, he sent Masson to Quebec in 1797 – a decision that placed both the plant hunter and the plants he collected in peril.

---

[49] John Gascoigne has recounted how, in March 1793, Banks was astounded by one French correspondent who refused to visit him because of the two countries' political differences. Gascoigne, *Science in the Service of Empire*, p. 155.

[50] PSJB, 5.13.62, Francis Masson (Cape of Good Hope) to Joseph Banks, 21 June 1794.

[51] PSJB, 5.13.63, Francis Masson (Cape of Good Hope) to Joseph Banks, 9 October 1794.

[52] PSJB, 5.13.56, Francis Masson (Cape of Good Hope) to Joseph Banks, 25 February 1793. Banks' annotation is dated 9 July 1795.

[53] PSJB, 5.13.63, Francis Masson (Cape of Good Hope) to Joseph Banks, 9 October 1794. See also: Rediker, *Devil and the Deep Blue Sea*, p. 35.

[54] PSJB, 5.13.62, Francis Masson (Cape of Good Hope) to Joseph Banks, 21 June 1794.

Masson's ship was attacked by French privateers during his passage across the Atlantic, and his itinerary through Canada was to a large extent determined by the need to avoid areas of fighting. As Canadian ports were unsafe, Masson was obliged to send duplicates of his specimens overland to New York – a lengthy journey that damaged and weakened the plants even before they had been laded aboard ship.[55] Nevertheless, in 1796, Banks had assured Muséum botanist Antoine-Laurent de Jussieu that 'whatever the vicissitudes of war,...the sciences, and those who, like yourself, allow themselves to search out their sublime vistas, will always hold the place closest to [my] heart'. In 1800, Banks also famously helped negotiate the liberation of French geologist Déodat de Dolomieu from imprisonment by the King of Naples.[56]

British cosmopolitan attitudes shifted more slowly than those of the French, but by the early nineteenth century the conceptual alignment between science and the state was gaining strength. Even Banks' interactions with his French counterparts were eventually inflected by rising nationalism, and he found himself maintaining a fine balance between his universalist commitments and his loyalty to his country. Writing just before the peace negotiations for the Treaty of London in October 1801, Banks added a postscript to a letter regarding Dolomieu, reminding the French recipient that 'we in England, sir, are as firmly attached to Royal Government as you can be to Republican'.[57] By 1814, even Banks acknowledged that war conditions had made it almost impossible to obtain plants from foreign countries. Nevertheless, he noted optimistically that 'A treaty with France would reopen its possibility'.[58]

The constraints exerted by the Napoleonic Wars reduced the opportunities available to scholars for international exchange. Public antipathy towards an enemy nation also meant that cosmopolitan behaviour was no longer promoted overtly as an ideal. Indeed, the French publicised and celebrated seizures of foreign collections. The *Magasin Encyclopédique*, for example, informed its readers in detail not only about the arrival in Paris of the Dutch Stathouder's cabinet in 1795, but also about smaller prizes, such as 'a box of seeds from an English ship of benefit to the Bordeaux Garden' in 1798, and plants from an English ship in 1799.[59]

[55] PSJB, 5.13.66, Copy [by William Cartlich] of a letter from Francis Masson (Montreal) to Joseph Banks, 5 November 1799; PSJB, 5.13.68, Francis Masson (Quebec) to Joseph Banks, 1 November 1800. See also: Jarrell, 'Masson, Francis'.

[56] Daston, 'Nationalism', p. 100; Gascoigne, *Science in the Service of Empire*, p. 156.

[57] Quoted in Daston, 'Nationalism', p. 100.

[58] Summary of letter from Joseph Banks to William Aiton Jr, 7 June 1814, in Banks, *Banks Letters*, p. 11.

[59] *Magasin Encyclopédique*, vol. 2 (1795), p. 419 (Stathouder); vol. 2 (1798), pp. 545–546 (box of seeds); vol. 1 (1799), p. 383 (plants). On the legality of these seizures, and

Although cosmopolitan behaviours became less visible, they nevertheless continued to feature in scholars' private dealings with one another. The *Magasin Encyclopédique* reported exchanges as well as confiscations: in 1803, it noted that Lee and Kennedy had gifted to France 100 very rare species of South African heather; in return, they were given four boxes of plants selected from those brought back from the Antilles by the Baudin expedition.[60] The elevated aspiration of maintaining a scholarly community in the service of 'humanity' continued to weave a thread through scientific practice in the early nineteenth century.

The French Revolution and then Napoleon inaugurated new, less cooperative attitudes towards science and the sharing of data, which were underpinned by nationalistic arguments about the relationship between science and nation. In spite of the new ethos of Napoleonic science, however, many scholars remained remarkably committed to the circulation of information. What is also remarkable is that commercial nurseries in France as well as Britain also continued to uphold transnational communication and exchange. What changed in France was the way in which the outcomes of such exchanges were communicated to the public.

The French nurseries were also noteworthy during the Revolution for the different strategies they used to ensure their survival. Vilmorin and Cels, in particular, drew upon approaches they had already been developing prior to 1789, deepening their existing connections within intellectual communities that evolved into government committees explicitly formed to serve the new nation. In this respect, they were also lucky – the clear utilitarian output of natural history meant that it retained respectability even during the worst excesses of the Terror. That association also helped them to distance themselves from what became a toxic cultural connection between flower collecting and luxury.

Were Vilmorin and Cels exceptional compared to other merchants and gardeners? The fortunes of their counterparts who had performed similar roles prior to the Revolution were tremendously varied, but this did not lead to the demise of gardening overall. Thomas Blaikie – the plant hunter discussed in Chapter 4, who had immigrated to France in 1776 to make gardens for the French aristocracy, including the Duc d'Artois – survived the Revolution, and continued to cultivate plants and

the question of whether they could be counted as spoils of war, see: Lipkowitz, 'Seized natural history collections'. See also: Lacour, *La République Naturaliste*, pt. 1.

[60] *Magasin Encyclopédique*, vol. 4 (1803), p. 258. On Baudin, start with: Harrison, 'Projections of the revolutionary nation'; Horner, *The French Reconnaissance*; Jangoux, 'L'expédition aux Antilles', p. 43; Sankey *et al.*, 'The Baudin expedition'; Starbuck, *Baudin*.

share specimens and information with his French and British correspondents. The cultivation of ornamental plants regained its cultural place after the Terror, and especially under Napoleon. Empress Josephine, after all, famously cultivated the gardens at Malmaison from 1800 until her death in 1814.

Studying international botanical exchanges from the perspective of commercial nurseries and plant hunters nuances existing accounts of cosmopolitan science. The latter have generally taken an approach that focuses on institutions such as the Jardin du Roi, or on important individuals such as Joseph Banks. While these institutions (and quasi-institutions, for Banks) certainly offered a framework for cosmopolitan science by promoting the ideal and negotiating international exchanges, and by commissioning and directing plant hunters, they controlled neither it, nor international plant exchanges, completely. Plant hunters and commercial nurseries were also implicated in the articulation of cosmopolitan ideals and practices on a day-to-day basis. Periods of warfare put pressure on these ideals and practices, especially during the Revolutionary and Napoleonic Wars. But scholars' and traders' vested interests in science or in commercial profit (or in both science *and* profit) meant that plants continued to be exchanged on a global scale whenever it was practically possible. This was in spite of the fact that public rhetoric about science became more closely aligned with national objectives, first in revolutionary France and then in Britain by the early nineteenth century.

Conclusion
Commerce and Cultivation

This book has traced the lives and activities of British and French upper-end plant traders, using their experiences to explore the social and cultural worlds that surrounded eighteenth- and early nineteenth-century botany and horticulture. Merchants like Lewis Kennedy and James Lee, and Adélaïde d'Andrieux and Philippe-Victoire Lévêque de Vilmorin, primarily sold their wares to consumers keen to obtain ornamental flowers, shrubs and trees. They were set apart from the majority of plant traders by their ability to obtain and cultivate rare, often exotic, specimens. The quantity of new plants arriving in Europe greatly increased over the second half of the eighteenth century, and many of these specimens required careful cultivation prior to sale. Naturalising tender exotics in new climates demanded that the men and women who worked within plant nurseries developed a sophisticated practical expertise.

The exotic specimens may have been demanding, but they brought with them exciting new social opportunities and commercial rewards for the merchants. Key to the traders' success was their ability to cultivate social connections, not only with their customers but also – and most especially – with scholars of botany. The merchants supplied specimens and contributed practical observations about the growth and behaviour of the plants. The upper-end traders understood botany and (even if they did not claim the title of 'botanist') helped to solve the puzzles that exercised eighteenth-century botanists – about taxonomy and nomenclature, for example. Many of the merchants discussed here capitalised so effectively on this hybrid expertise that their businesses lasted well into the nineteenth century, and in one case right to the present day.

Viewing Enlightenment botany from the perspectives of the plant traders invites us to rethink several received ideas about the cultural and social history of science in the late eighteenth and early nineteenth centuries. These concern, broadly, the links between commercial culture and public science, and the structure and composition of Enlightenment networks.

189

Attention to the clientele on whose custom the traders depended led to a reassessment of the cultural framework in which public participation in science was often situated. The traders marketed their specimens to a public that primarily comprised the middling and upper ranks. Unlike many of their counterparts, however, they did not simply sell plants as *plants*. Instead, they emphasised the potential scientific interest that their rarest specimens could attract. The plant traders primarily engaged with the figure of the amateur; their advertisements and other publications gave clear indications of the meaning of amateur botany, and showed how notions about amateur scholarship were refracted through assumptions about social status and gender. Amateur scholarship meant a refined, tasteful engagement with learning and demanded an education of the vision that was closely associated with connoisseurship of the fine arts.

The study of botany was thus situated within a particular aesthetic and social framework. Amateurship was primarily considered a masculine attribute, and was normally associated with the upper ranks. These discursive associations had two significant effects on the cultures of botany in the decades around 1800. On the one hand, the connection between amateur botany, refinement and good taste helped to legitimise the participation of elite or middle-ranking men in floriculture, which was otherwise associated with effeminacy and the lower social ranks. On the other hand, the notion that amateur botany was an aspect of elite male culture complicated the participation of women of all ranks, and of middle-ranking men who did not identify with that culture. Amateur culture, thus, was responsible for defining the ways in which men and women participated (or did not participate) in botanical networks.

In contrast to the received idea that Enlightenment botany was dominated by taxonomy and justified through its potential applications to agriculture, medicine and industry, the amateur botanists discussed here were more concerned with the cultivation of good taste, with obtaining new ornamental specimens and with growing such specimens as live plant collections. Horticulture, as a practical form of natural knowledge, was upheld as part of botanical study. Enlightenment botany fostered a host of scholarly and commercial activities that strayed far beyond a concern with taxonomy. These activities were replicated and reinforced by professional gardeners and commercial nurseries.

In practice, of course, all sorts of men and women from the middling and upper ranks, aided by the resources offered by the expanding scientific marketplace, took up botany. The people who engaged with the scientific study of plants ranged from those with an extensive expertise to those who studied it fleetingly at most. Notions of amateurship were very evident within cultures of botany, and many people defined their engagement in science through or against this cultural model. Male and female amateurs applied their knowledge in ways that could be, at one extreme, straightforwardly utilitarian, or, at the other, beguilingly creative.

The attention paid to public science also led us to reconsider the nature of Enlightenment scholarship, especially in terms of its transnational nature, social composition and longevity. France and Britain were not identical in terms of the ways in which scientific research and practice was organised, but there were strong parallels between the two countries. Botanical scholars at royal institutions such as the Jardin du Roi, Kew Gardens, the Académie des Sciences and the Royal Society interacted extensively with a network of non-institutional actors composed of gardeners and plant merchants. The latter played a major role in supporting scientific research. They also formed strong intellectual links with counterparts across the Channel. Enlightenment botany was transnational in composition and outward looking in ethos.

The extended examination of merchants' and plant hunters' experiences has thus revised existing assumptions about how non-elite scholars participated in the Republic of Letters. The plant traders and gardeners who have peopled the pages of this book were not members of the Republic of Letters. Their skills and expertise, however, meant that members of this community were willing to engage with them, sharing information and specimens. James Lee, Thomas Blaikie, Philippe-Victoire Lévêque de Vilmorin and their counterparts came to be part of a socially diverse network of knowledgeable and proficient practitioners that supported Enlightenment scholarship. The scholarly commerce in which they participated underpinned in significant ways the making of new natural knowledge.

Reflecting the ethos of the Republic of Letters, most plant merchants and plant hunters subscribed to the cosmopolitan ideal that promoted unpartisan behaviour in order to increase knowledge overall. In spite of some clear instances of state direction of botanical research (for example, colonial botanical schemes intended to serve imperial aspirations), botany emerged primarily in a transnational context that fostered cosmopolitan behaviours. Those connections continued through most of the late eighteenth and early nineteenth centuries: specimens and information were shared across national borders with little regard to the political differences that might exist between interlocutors. The openness that characterised botanical communication and exchange meant that, although national inflections are evident at particular points within a network, it is rare to find stark differences and contrasts among individuals in different nations.

The period between 1789 and 1815 is generally presented in terms of the rise of radicalism and conservatism, and as characterised by economic retrenchment and the growth of nationalism, in Britain as well as in France. In extending the enquiry into the first decades of the nineteenth century, this book has upheld a thesis about the continuity of Enlightenment scholarly culture, in spite of the efforts to nationalise science under Napoleon. Plant traders and their scholarly associates continued to exchange specimens and information whenever they could, even during the Napoleonic Wars. To the best of their ability, they maintained connections with their counterparts across Europe.

How did the figure of the upper-end plant trader or gardener evolve in the nineteenth century, especially with regards to changes in the provision of scientific education and the shifting social profile of the scientific practitioner? The social and scientific contexts in which the science of plants was studied altered over the course of the nineteenth century, and the gap between professional and amateur gradually widened.[1] From the 1860s onwards, pejorative references to 'amateurs' of science became increasingly common in both France and Britain. Non-professional scientists risked derision as 'dabblers' or 'superficial' students of science. New institutions were established that formalised the curriculum for students of botany, and clearer differences then emerged between 'low' scientific culture and the research undertaken by qualified professionals. By the twentieth century, formal training programmes and specific qualifications ultimately came to demarcate the scientific community.[2]

Public interest in studying botany and creating ornamental gardens increased greatly over the nineteenth century. The expansion of consumer society in both Britain and France, and the resulting growth of anonymity within the marketplace, changed the relationship between nurseries and consumers, and heralded a general decline in the practices that had structured polite consumption in the eighteenth century. Shopping became increasingly capitalistic, characterised by the use of fixed prices, a reluctance to accept credit, greater anonymity among sellers and consumers, and even greater social diversity among

---

[1] Alberti, 'Amateurs and professionals'; Gay and Barrett, 'Should the cobbler stick to his last?'; Nye, 'On gentlemen'.

[2] The exact chronology of the changes within the scientific community differed between Britain and France. See: Sheets-Pyenson, 'Popular science periodicals'; 'Amateur' in OED Online, http://www.oed.com/view/Entry/6041 [accessed 26 August 2016]; 'Amateur' in *Le Trésor de la langue française informatisé*, http://atilf.atilf.fr/ [accessed 26 August 2016]. On the nineteenth-century transitions, see the contributions to Fyfe and Lightman, *Science in the Marketplace*, especially Secord, 'Scientific conversation'. See also: Fox, *The Savant and the State*; Secord, 'Corresponding interests'; Shteir, *Cultivating Women*, pp. 31–32.

clientele.[3] The move towards professionalisation in the sciences and the shifts in consumer culture did not immediately diminish the standing of upper-end plant traders such as Lee and Vilmorin. It seems, however, that both these changes altered how plant traders competed with one another and how they constructed their reputations. In particular, they developed a concept of ownership of their productions that was strikingly modern. Indicated first through the names given to new cultivars, and ultimately by attempts to identify plant varieties genetically, breeders increasingly sought to claim intellectual property over new productions.[4] Such assertions were quite at odds with the values expressed by their predecessors, who had demonstrated a firm commitment to the interests of a wider scientific and mercantile community.

Returning to the decades around 1800, the upper-end plant traders' ability to communicate with people of different social and national backgrounds was fundamental to their success. They were able to do this primarily by following the protocols set out by the culture of politeness that had pervaded eighteenth-century middle- and upper-ranking culture, and which continued into the early nineteenth century. The traders' interpretations of polite commerce proved attractive to customers, and so they used politeness, along with promotion of their amateurship and their participation in scholarly networks, as a means of marketing their wares. Cultural influences are, of course, rarely if ever unidirectional. The plant traders then found themselves negotiating the conventions that defined polite consumption, as well as the mores of Enlightenment science.

On 3 October 1789, head gardener André Thouin was at his desk at the Jardin du Roi in Paris, writing to Hippolyte Nectoux, his counterpart at the Jardin du Roi in Saint Domingue. In his letter, Thouin proposed that he would identify any unknown specimens that Nectoux could send. In this way, he and Nectoux would reciprocally keep each other abreast of the latest Caribbean discoveries. Thouin's letter testifies to the growing global interconnectedness felt by botanists and gardeners in the late eighteenth century, and to the rapid expansion of knowledge about the world's flora that resulted from such connections. But Thouin offered his correspondent more than knowledge from the cutting edge of science. He continued by asking Nectoux whether he would like to 'make an Object of speculation of the dry plants':

---

[3] Berry, 'Polite consumption'; Finn, *Character of Credit*.
[4] Bonneuil, 'Mendelism; Gaudillière *et al.*, *Living Properties*'; Kevles, 'Patents, protections and privileges'.

I offer my [services] to you to find a ... [market] among our Botanists and our Amateurs in England, in Holland, and in some parts of Germany. Samples of well-conserved dry plants ... sell for between twenty-five and fifty *livres* per hundred different species ... according to the rarity of the objects or their individual merit.[5]

Thouin's proposal was symptomatic of the longer-term developments that have been discussed in the course of this book. By the final decade of the eighteenth century, it was clear that botany was a lucrative science. Even the head gardener of France's top botanical institution saw fit to dabble in trade.

---

[5] BNF, NAF 9545, ff. 18–19, André Thouin (Paris) to Hippolyte Nectoux (Jardin du Roi, Saint Domingue), 3 October 1789.

Bibliography

ARCHIVAL DOCUMENTS

ARCHIVES DE PARIS

D.5B6 1 à 3000, Commerçants faillis.

ARCHIVES DÉPARTEMENTALES, LOIRE-ATLANTIQUE

L626, Observations relatives à l'établissement projeté d'un jardin de Botanique en la commune de Nantes, 15 Gérminal Year 3 [4 April 1795].

ARCHIVES NATIONALES DE FRANCE, PARIS

399/AP/97, Chartrier de Malesherbes.
399/AP/98, Chartrier de Malesherbes.
399/AP/99, Chartrier de Malesherbes.
399/AP/100, Chartrier de Malesherbes.
399/AP/101, Chartrier de Malesherbes.
AJ/15/149, Dépenses du Jardin du Roi.
AJ/15/511, Envois de graines, plantes, minéraux etc de pays étrangers au Jardin du Roi.
F/7/3580, Police Générale: Demandes des Passeportes J-Z 1793–1818.
F/10/371, M. Féburier, Observations sur les pépinières du Gouvernement, 2 January 1808.
O/1/2113A, Michaux à l'Amérique Septentrionale.

Minutier Central
ET/XLIV/0765, Inventaire après décès de Jacques Martin Cels, 27 June 1808.
RE/LXXV/22 and ET/LXXV/1132, Inventaire après décès d'Adélaïde d'Andrieux, 11–12 March 1836.

BIBLIOTHÈQUE NATIONALE DE FRANCE, PARIS

NAF 2757, Agronomes – correspondance.
NAF 2758, Agronomes – correspondance.
NAF 9545, Nectoux – correspondance et papiers.

BODLEIAN LIBRARY, OXFORD

The John Johnson Collection of Printed Ephemera, www.bodleian.ox.ac.uk/
johnson [accessed 24 August 2016]
Ms Douce c. 11, Francis Douce, Notes on Antiquarian Subjects.
Ms Douce d. 39, Letters to Francis Douce from Richard Twiss.
Ms Eng. Let. c.229, Tunstall-Constable Correspondence.

BRITISH LIBRARY, LONDON

Add. Ms. 8094, Joseph Banks, Correspondence.
Add. Ms. 8095, Joseph Banks, Correspondence.
Add. Ms. 8096, Joseph Banks, Correspondence.
Add. Ms. 8097, Joseph Banks, Correspondence.
Add. Ms. 8098, Joseph Banks, Correspondence.
Add. Ms. 18565, Kaye Notebooks, vol. XVI.
Add. Ms 28534, Emanuel Mendes da Costa, Correspondence.
Add. Ms. 28536, Emanuel Mendes da Costa, Correspondence.
Add. Ms. 28540, Emanuel Mendes da Costa, Correspondence.
Add. Ms. 29533, Letters from naturalists. Presented by J.E. Gray.
Add. Ms. 33540, Bentham Papers. Correspondence vol. IV. 1784–1788.
Add. Ms. 33977, Joseph Banks correspondence, vol. I 1765–1784.
Add. Ms. 33980, Joseph Banks correspondence, vol. IV 1795–1801.
Add. Ms. 35057, Miscellaneous Letters, 1693–1868. Presented by S. G.
Perceval.
Add. Ms. 47237, Lieven Papers.

BRITISH MUSEUM, LONDON

Banks and Heal Collections of Printed Ephemera.

CUMBRIA RECORD OFFICE, CARLISLE

D/HC/1/81, Catherine Mary Howard, Reminiscences for my Children (1831).
D/Sen/5/5/1/9, Thomas Dixon (Netherby, Cumberland) to John Johnston
(Netherhall, Cumberland), 16 December 1802.

LANCASHIRE RECORD OFFICE, PRESTON

DDWH/4/138, I. S. Howson (Sunderland?) to Tom Whittaker (Macclesfield
Grammar School), ?1831.

LIBRARY OF CONGRESS, WASHINGTON, DC

Jefferson Papers: Coolidge Collection.

THE LINNAEAN SOCIETY OF LONDON

The Linnaean Correspondence Online, http://linnaeus.c18.net.
L4741; L5155; L5238; L5239; L5601.

MUSÉUM NATIONAL D'HISTOIRE NATURELLE, PARIS

Ms 47, Joseph Martin, Documents divers concernant les Îles de France et de
Bourbon.

Ms 307, Végétaux envoyés au Sultan Tippo [sic] Zaib par le Jardin du Roi 1788.

Ms 314, État de la Correspondance d'André Thouin.

Ms 318, Dossier XV, Vilmorin. Mémoire sur les semis d'arbres et d'arbustes
(n.d.) [before 1792].

Ms 357, Papiers provenant de Louis-Guillaume Le Monnier (1717–1799) et en
partie de sa main.

Ms 1934/XXX, André Thouin, Mémoire sur le Jardin du Roi, October 1788.

Ms 1945, Expédition d'Aristide Aubert Dupetit-Thouars pour la Recherche de
La Pérouse, 1792.

THE NATIONAL ARCHIVES, LONDON

Prob 11/1241, Will of Anna Blackburne.

NATURAL HISTORY MUSEUM, LONDON

Banks Collection.
Botany Library, Banks Correspondence.
Dawson Turner Copies.
H. B. Carter Transcripts.

NOTTINGHAM TRENT UNIVERSITY

Sir Joseph Banks Archive Project.

OLD BAILEY PROCEEDINGS ONLINE (WWW.OLDBAILEYONLINE.ORG, VERSION 7.2, 11 AUGUST 2016)

16 September 1795, trial of Charles Fairfield (t17950916-73).

ROYAL BOTANIC GARDENS, KEW

Banks Collection.

SOCIÉTÉ VILMORIN, LA MENITRÉ, MAINE ET LOIRE

Boîte: Vilmorin – Vieux Documents.

STATE LIBRARY OF NEW SOUTH WALES, AUSTRALIA

The Papers of Sir Joseph Banks, www2.sl.nsw.gov.au/banks/.

WARRINGTON LIBRARY ARCHIVES

RWA Wp 82182, 'Orford Hall List of Natural History Specimens, Portraits etc', by Warrington Museum and Art Gallery, 1914.

WARWICK COUNTY RECORD OFFICE

CR 2017/TP 408, Thomas Pennant, Papers.

YALE UNIVERSITY, LEWIS WALPOLE LIBRARY

Eshott Papers Mss 2, Box 24, Records relating to Eshott House, c. 1758–1839.

NEWSPAPERS AND PERIODICALS

Avant-Coureur.
The Connoisseur.
Feuille du Cultivateur, Rédigée par les CC. Dubois, Lefebvre et Parmentier.
Gardener's Magazine.
Gentleman's Magazine.
Journal de Paris.
Magasin Encyclopédique, ou journal des sciences, des lettres et des arts.
Mémoires d'Agriculture, d'économie rurale et domestique.
Procès-Verbaux du Comité d'Agriculture et des Arts.
St James's Chronicle.

PUBLISHED PRIMARY SOURCES

Abercrombie, John, *The Complete Kitchen Gardener, and Hot Bed Forcer; with the Thorough Practical Management of Hot-Houses, Fire-Walls and Forcing-Houses* (London: J. Stockdale, 1789).
   The Garden Vade-Mecum, or Compendium of General Gardening (Dublin, 1790).
Aikin, John, *A Description of the Country from Thirty to Forty Miles Round Manchester* (London: J. Stockdale, 1795).
Aiton, William, *Hortus Kewensis; or, a Catalogue of the Plants Cultivated in the Royal Botanic Garden at Kew* (3 vols, London: George Nicol, 1789).
Andrews, Henry, *The Botanists' Repository for New and Rare Plants, Containing Coloured Figures of Such Plants as have not Hitherto Appeared in any Similar Publication* (10 vols, London: T. Bensley, 1797).
Andrieux [Adélaïde d'] and [Pierre-Victoire Lévêque de] Vilmorin, *Catalogue des plantes, arbres, arbrisseaux, et arbustes, dont on trouve des graines, des bulbes, & du plant, chez les sieurs Andrieux et Vilmorin, Marchands Grainiers-Fleuristes & Botanistes du Roi, & Pépiniéristes* (Paris, 1778).

Andrieux, [Pierre d'], *Catalogue de toutes sortes de graines, fleurs, oignons de fleurs &c.* (Paris, 1760).

Catalogue raisonné des plantes, arbres et arbustes dont on trouver des Graines, des Bulbes et du Plant chez le sieur Andrieux (Paris, 1771).

Anon. *Le Bon Jardinier. Almanach pour l'année 1773. Contenant une idée générale des quatre sortes de Jardins, les règles pour les cultiver, la manière de les planter, et celle d'élever les plus belles Fleurs. Nouvelle édition* (Paris: Guillyn, 1773).

'Jardinage', in Diderot and D'Alembert, *Enclyclopédie*, vol. 8 (1765), p. 460.

Jardinier Portatif, ou la Culture des Quatre Classes de Jardins, et de l'Educaton des Fleurs (Liège: F.J. Desoer, 1772).

'Observations on such nutritive Vegetables as may be substituted in the Place of ordinary Food in Times of Scarcity', *Gentleman's Magazine* 53 (June 1783): 517.

'Persons of quality proved to be traders', *St James's Chronicle*, no. 51 (9 July 1761): 182.

Traité de la Culture de Différentes Fleurs (1765).

Anon. [Daubenton?], 'Histoire naturelle', in Diderot and D'Alembert, *Enclyclopédie*, vol. 8 (1765), pp. 226–228.

Anon. [de Grace?], *Le Bon Jardinier, Almanach Pour l'Année 1768. Contenant une idée générale des quatre sortes de Jardins, les regles pour les cultiver, la maniere [sic] de les planter & celle d'élever les plus belles Fleurs* (Paris: Guillyn, 1768).

Ash, John, *The New and Complete Dictionary of the English Language* (London, 1775).

Banks, Joseph, *The Banks Letters. A Calendar of the Manuscript Correspondence of Sir Joseph Banks Preserved in the British Museum (Natural History) and Other Collections in Great Britain*, ed. Warren R. Dawson (London: British Museum, 1958).

The Endeavour Journal of Joseph Banks, 1768–1771, ed. J. C. Beaglehole (2 vols, Sydney: Angus and Robertson, 1962).

The Letters of Sir Joseph Banks. A Selection, 1768–1820, ed. Neil Chambers (London: Imperial College Press, 2000).

Barbé-Marbois, François, *Our Revolutionary Forefathers: The Letters of François, Marquis de Barbé-Marbois*, trans. and ed. Eugene Parker Chase (New York: Duffield & co., 1929).

Barr, James Smith, *Barr's Buffon. Buffon's Natural History: Containing a Theory of the Earth, a General History of Man, of the Brute Creation, and of Vegetables, Minerals, &c.*, 10 vols (London: J. S. Barr, 1792).

Barrin de La Galissonière, Roland-Michel and Henri-Louis Duhamel du Monceau, *Avis pour le transport par mer des arbres, des plantes vivaces, des semences et de diverses autres curiosités d'histoire naturelle* (2nd edn, Paris: Imprimerie Royale, 1753).

Barrington, Daines, *On the Progress of Gardening. In a Letter from the Hon. Daines Barrington to the Rev. Mr. Norris* (London, 1785).

Bartram, John and Humphrey Marshall, *Memorials of John Bartram and Humphrey Marshall*, ed. William Darlington (New York and London: Hafner Publishing Company, 1967 [1849]).

Blaikie, Thomas, *Diary of a Scotch Gardener at the French Court at the End of the Eighteenth Century*, ed. with intro. Francis Birrell (London: George Routledge and Sons, 1931).

'Foreign notices', *The Gardener's Magazine*, vol. 3 (1828): 207.

Sur les terres d'un jardinier. Journal de voyages 1775–1792, trans. Janine Barrier, ed. Janine Barrier and Monique Mosser (Paris: Les Éditions de l'Imprimeur, 1997).

Bradley, Richard, *A General Treatise on Husbandry and Gardening* (London, 1726).

Buffon, George-Louis Leclerc, comte de, 'Initial discourse: on the manner of studying and expounding natural history', trans. John Lyon, in Lyon and Sloan, *From Natural History to the History of Nature*.

Calonne, Louis-François de, *Essais d'agriculture en forme d'entretiens... par un cultivateur à Vitry-sur-Seine* (Paris, 1778).

Coypel, Charles-Antoine, *La Curiosimanie*, eds. Florence Buttay, Laurence Macé and Nathalie Volle (Librairie des Musées, 2014).

Curtis, William, *Proposals for Opening by Subscription, A Botanic Garden* (London, 1778).

Cushing, John, *The Exotic Gardener* (2nd edn, London: G. and W. Nicol, 1814).

Darwin, Erasmus, *The Botanic Garden, Part II. Containing the Loves of the Plants. A Poem, with Philosophical Notes* (London: J. Johnson, 1789).

de Grace, *Le Bon Jardinier, Almanach pour l'année M.D.CC.LXXXV... Nouvelle édition avec suppléments* (Paris: Eugène Onfroy, 1785).

Delany, Mary, *The Autobiography and Correspondence of Mary Granville, Mrs Delany*, ed. Lady Llanover, ser. 1 (London, 1861).

Dicksons &co., *A Catalogue of Hot-House, Green-House and Hardy Plants; Flowering and Evergreen Shrubs, Fruit and Forest Trees* (Edinburgh, 1792).

Diderot, Denis, 'Explication détaillée du système des connaissances humaines', in Diderot and D'Alembert, *Encyclopédie*, vol. 1 (1751), pp. xlvii–li.

'Salon de 1767. Adressé à mon ami M. Grimm', in Versini, Diderot. Œuvres, vol. 4, pp. 518–529.

and Jean le Rond d'Alembert (eds.), *Encyclopédie, ou Dictionnaire Raisonné des Sciences, des Arts et des Métiers* (35 vols, facsim. repr. Stuttgard-Bad Cannstatt: Friedrich Frommann Verlag [Günther Holzboog], 1966 [Paris: Briasson, David, Le Breton and Durand, 1751–1765]).

Dirom, Alexander, *A Narrative of the Campaign in India, which Terminated the War with Tippoo Sultan in 1792* (London: W. Bulmer, 1793).

Dumont de Courset, George Louis Marie, *Le botaniste cultivateur, ou Description, culture et usages de la plus grande partie des plantes étrangères, naturalisées et indigènes, cultivées en France et en Angleterre, rangées suivant la méthode de Jussieu* (5 vols, Paris: J. J. Fuchs, 1802–05).

Edgeworth, Maria, *Letters for Literary Ladies*, ed. Claire Connolly (London and Vermont: Everyman, 1993 [1798]).

Ellis, John, *Directions for Bringing Over Seeds and Plants, from the East-Indies and Other Distant Countries in A State of Vegetation* (London: L. Davis, Printer to the Royal Society, 1770).

Faulkner, Thomas, *The History and Antiquities of the Parish of Hammersmith, Interspersed with Biographical Notices of Illustratious and Eminent Persons...* (London: Nichols & Son, J. Weale, E. Page, T.S. Rayner and Simpkin, Marshall &co., 1839).

Galpine, John Kingston, *The Georgian Garden. An Eighteenth-Century Nurseryman's Catalogue*, ed. John H. Harvey (Dorset: The Dovecote Press, 1983).

Gerbaux, Fernand and Charles Schmidt (eds.), *Procès-Verbaux des Comités d'Agriculture et de Commerce de la Constituante de la Législative et de la Convention* (4 vols, Paris: Imprimerie Nationale, 1908).

Gillray, James, *The Great South Sea Caterpillar, Transform'd into a Bath Butterfly* (London, 1795).

Gordon, Dermer and Thomson, *A Catalogue of Trees, Shrubs, Plants, Flower Roots, Seeds, &c.* (London, 1783?).

Goring and Wright, *A Catalogue of Flowers, Plants, Trees, &c.* (London: Stephen Couchman, 1798).

Haller, Albrecht, *Historia Stirpium Indigenarum Helvetiæ Inchoata* (Berne, 1768). *Nomenclator ex Historia plantarum indigenarum Helvetiæ excerptus* (Berne, 1769).

Hill, John, *Exotic Botany Illustrated in Thirty-Five Figures of Curious and Elegant Plants: Explaining the Sexual System; and Tending to Give Some New Lights into the Vegetable Philosophy* (1759).

Hurtaut, Pierre-Thomas-Nicolas and L. de Magny, *Dictionnaire historique de la ville de Paris et de ses environs* (4 vols, Paris: Moutard, 1789).

Jacson, Maria Elizabeth, *Botanical Dialogues between Hortensia and her Four Children, Charles, Harriet, Juliette and Henry. Designed for the use of schools, by a lady* (London: J. Johnson, 1797).

Johnson, George W., *A History of English Gardening, Chronological, Biographical, Literary and Critical. Tracing the Progress of the art in this country from the Invasion of the Romans to the present time* (London: Baldwin &co., Wright, Ridgway and Wicks, 1829).

Kennedy, Lewis and James Lee, *Catalogue of Plants and Seeds, Sold by Kennedy and Lee, Nursery and Seedsmen, At the Vineyard, Hammersmith* (London: S. Hooper, 1774).

Kenrick, William and John Murdoch, *Natural History of Animals, Vegetables, and Minerals; with the Theory of the Earth in General*, 6 vols (1775).

Landois, Paul, 'Amateur', in Diderot and D'Alembert, *Encyclopédie*, vol. 1 (1751), p. 317.
'Connoisseur', in Diderot and D'Alembert, *Encyclopédie*, vol. 3 (1753), p. 898.

La Pérouse, Jean-François de Galaup de, *The Journal of Jean-François de Galaup de la Pérouse 1785–1788*, trans. and ed. John Dunmore (2 vols, London: The Hakluyt Society, 1994).

Le Berryais, René, *Traité des jardins, ou le nouveau de La Quintinye* (3rd edn, Paris: Belin, 1789).

Lebreton, François, *Manuel de Botanique, à l'usage des amateurs et des voyageurs* (Paris: Prault, 1787).

Lee, James, *An Introduction to Botany. Containing an Explanation of the Theory of that Science, and an Interpretation of its Technical Terms* (London: J. and R. Tonson, 1760).

An Introduction to Botany. Containing an Explanation of the Theory of that Science, and an Interpretation of its Technical Terms (2nd edn, London: J. and R. Tonson, 1765).

An Introduction to Botany. Containing...An Appendix and Glossary (5th edn, London: S. Crowder, C. Dilly et al., 1794).

Introduction to the Science of Botany ('4th edn' [actually the 10th edn], London: C. and J. Rivington, Wilkie and Robinson, J. Walker et al., 1810).

Linné, Carl, *A System of Vegetables... Translated from the Thirteenth Edition... By a Botanical Society at Lichfield* (Lichfield and London: John Jackson, 1783).

Loddiges and Sons, *The Botanical Cabinet. Consisting of Coloured Delineations of Plants from All Countries. With a Short Account of Each, Directions for Management &c &c* (6 vols, London: John & Arthur Arch, John Hatchard, C. Loddiges & Sons, G. Cooke, 1817).

Luker and Smith, *A Catalogue of Trees, Shrubs, Greenhouse Plants, Seeds, and Bulbous Roots* (1783).

Maddox, James, 'A catalogue of gooseberry trees, raised from seed in Lancashire, to the year 1780', in Weston, *Supplement to the English Flora*, pp. 115–118.

Malcolm, William, *A Catalogue of Hot-House and Green-House Plants* (London, 1778).

Mariette, P.-J., *Traité des pierres gravées* (Paris, 1750).

Mawe, Thomas and John Abercrombie, *Every Man His Own Gardener* (18th edn, London, 1805).

Miller, Philip, *The Gardener's Dictionary. Containing the Methods of Cultivating and Improving the Kitchen, Flower, Fruit and Pleasure Garden* (6th edn, London, 1752).

More, Hannah, *Strictures on the Modern System of Female Education* (3rd edn, London: T. Cadell and W. Davies, 1799).

Mortimer, John, *The Whole Art of Husbandry: or, The Way of Managing and Improving of Land. Being a Full Collection of What Hath Been Writ, either by Ancient or Modern Authors* (5th edn, London, 1721).

Murdoch, Alexander, 'Campbell, Archibald, third duke of Argyll (1682–1761)', in *Oxford Dictionary of National Biography* (Oxford University Press, 2004); online edn, October 2006, www.oxforddnb.com/view/article/4477 [accessed 22 April 2016].

Murith, Laurent-Joseph, *Le Guide du Botaniste qui Voyage dans le Valais, avec un Catalogue des Plantes de ce Pays et de ses Environs, Auquel on a joint les Lieux de Naissance et l'Époque de la Fleuraison pour Chaque Espèce* (Lausanne: Henri Vincent, 1810).

Murray, Lady Charlotte, *The British Garden* (Bath: S. Hazard, 1799).

Neal, Adam, *A Catalogue of the Plants in the Garden of John Blackburne, Esqr at Orford, Lancashire. Alphabetically Arranged According to the Linnæan System* (Warrington: William Eyres, 1779).

Neill, Patrick, *Journal of a Horticultural Tour. Through Some Parts of Flanders, Holland, and the North of France. In the Autumn of 1817* (Edinburgh: Bell & Bradfute; London: Longman, Hurst, Rees, Orme & Brown, 1823).

Parker, George, Earl of Macclesfield, *The Earl of Macclesfield's Speech in the House of Peers on Monday the 18th Day of March, 1750. At the Second Reading of the Bill, for Regulating the Commencement of the Year* (Dublin: George Faulkner, 1751).

Parkinson, John, *Paradisi in Sole Paradises Terrestris, or, A Garden of All Sorts of Pleasant Flowers which our English Ayre will Permitt to be Noursed up* (London, 1629).

Pennant, Thomas, *Arctic Zoology* (2 vols, London: Henry Hughes, 1784–85).

A Tour in Scotland, and Voyage to the Hebrides (2nd edn, London: Benjamin White, 1776).

Perfect, William and John Perfect, *A Catalogue of Forest-Trees, Fruit-Trees, Evergreen and Flowering Shrubs* (York: C. Etherington, 1777).

Pindar, Peter (pseud.), *Sir Joseph Banks and the Emperor of Morocco* (London, 1788).

Polwhele, Richard, *The Unsex'd Females. A Poem, addressed to the Author of The Pursuits of Literature* (2nd edn, New York: William Cobbett, 1800 [1798]).

Powell and Eddie, *North American Tree, Shrub and Plant Seeds, Imported and Sold* (c. 1764).

Remy, Pierre, *Catalogue raisonné des tableaux, estampes, coquilles ...* (1766).

Rousseau, Jean-Jacques, *Letters on the Elements of Botany: Addressed to a Lady*, trans. Thomas Martyn (2nd edn, London: B. White and Son, 1787 [1785]).

Lettres élémentaires sur la botanique, in Collection complète des œuvres de J. J. Rousseau, Citoyen de Genève, vol. 7 (Geneva, 1782).

and Chrétien-Guillaume de Lamoignon de Malesherbes, *Jean-Jacques Rousseau, Chrétien-Guillaume de Lamoignon de Malesherbes: Correspondence*, ed. Barbara de Negroni (Paris: Flammarion, 1991).

[and Thomas Martyn], *Lettres élémentaires sur la botanique*, in Œuvres complètes de J. J. Rousseau. Nouvelle édition, classée par ordre de matières, et ornée de quatre-vingt-dix gravurs, eds. Louis Sebastian Mercier, Pierre Philippe Félicien Le Tourneur, Gabriel Brizard and François Henri Stanislas de l'Aulnaye, vol. 5 (Paris: Poincot, 1789).

Shaftesbury, Antony Ashley Cooper, 3rd Earl of, *Soliloquy: or, Advice to an Author* (London: John Morphew, 1710).

Silvestre, Augustin-François, 'Notice biographique sur P.-V.-L. de Vilmorin', in *Séance publique de la Société d'Agriculture* (Paris: Société d'Agriculture, 26 Brumaire an XIV [17 November 1805]).

Smellie, William, *Natural History, General and Particular, by the Count de Buffon*, 8 vols (London: W. Creech, 1780).

Smith, James E., 'A review of the modern state of botany', *Supplement to the Fourth, Fifth and Sixth Editions of the Encyclopaedia Britannica* 2 (1824).

Smollet, Tobias, *The Expedition of Humphrey Clinker* (1771).

Society of Gardeners, *Catalogus Plantarum ... A Catalogue of Trees, Shrubs, Plants, and Flowers, Both Exotic and Domestic, Which are propagated for Sale in the Gardens near London* (London, 1730).

Stephens, E. B., 'Review: On the application of mangel wurzel and potatoes to the manufacture of sugar and spirits', *The National Magazine* 1 (1830): 229–335.

Swift, Jonathan, *Gulliver's Travels* (1726).

Telford, John and George Telford, *A Catalogue of Forest-Trees, Fruit-Trees, Evergreen and Flowering Shrubs* (York: A. Ward, 1775).

Thiéry, Luc-Vincent, *Guide des amateurs et des étrangers voyageurs à Paris* (2 vols, Paris: Hardouin et Gattey, 1787).

Thomson, James, *The Four Seasons, and other Poems* (London, 1735).

Thornton, Robert, 'Sketch of the life and writings of the late James Lee', in Lee, Introduction to the Science of Botany, pp. v–xix.

Twiss, Richard, *A Trip to Paris, in July and August, 1792* (London: William Lane and Mrs Harlow, 1793).

Ventenat, Étienne Pierre and Pierre-Joseph Redouté, *Description des plantes nouvelles et peu connues, cultivées dans le jardin de J.-M. Cels* (Paris, 1799).

Voltaire, François-Marie Arouet, Charles-Louis de Secondat, baron de La Brède et de Montesquieu, Denis Diderot and Jean Le Rond D'Alembert, 'Goût', in Diderot and D'Alembert, *Encyclopédie*, vol. 7 (1757), p. 761.

Wakefield, Gilbert, *Memoirs of the Life of Gilbert Wakefield* (London: J. Deighton, 1792).

Wakefield, Priscilla, *An Introduction to Botany, in a Series of Familiar Letters, with Illustrative Engravings* (Dublin: Thomas Burnside *et al.*, 1796).

   Reflections on the Present Condition of the Female Sex, with Suggestions for its Improvement (London: J. Johnson, 1798).

Webb, John, *A Catalogue of Seeds and Hardy Plants. With Instructions for Sowing and Planting* (London, 1760).

Weston, Richard, *Flora Anglicana . . . The English Flora: or, a catalogue of Trees, Shrubs, Plants and Fruits* (London: J. Millan; Robson and Co.; T. Carnan; E. and C. Dilly, 1775).

   Supplement to the English Flora (London: J. Millan, J. Robson, R. Faulder, T. Carnan and C. Dilly, 1780).

   Tracts on Practical Agriculture and Gardening (2nd edn, London: S. Hooper, 1773).

   Universal Botanist and Nurseryman (4 vols, London and York: J. Bell, G. Riley, J. Wheble and C. Etherington, 1770–77).

White, Gilbert, *The Natural History of Selbourne*, ed. Anne Secord (Oxford: Oxford University Press, 2013 [1789]).

Withering, William, *An Arrangement of British Plants; According to the Latest Improvements of the Linnæan System* (3rd edn, 4 vols, Birmingham, 1796).

   A Botanical Arrangement of all the Vegetables Growing in Great Britain (2 vols, Birmingham and London: T. Cadel, P. Elmsley and G. Robinson, 1776).

Wollstonecraft, Mary, *A Vindication of the Rights of Woman: With Strictures on Political and Moral Subjects*, ed. D. L. Macdonald and Kathleen Scherf (Toronto, ON: Broadview, 1997 [1792]).

Young, Arthur, *Arthur Young's Travels in France during the Years 1787, 1788, 1789*, ed. Miss Betham-Edwards (London: George Bell and Sons, 1900).

PUBLISHED SECONDARY SOURCES

Alberti, Samuel, 'Amateurs and professionals in one county: biology and natural history in late Victorian Yorkshire', *Journal of the History of Biology* 34(1) (2001): 115–147.

Albritton Jonsson, Fredrik, *Enlightenment's Frontier. The Scottish Highlands and the Origins of Environmentalism* (New Haven, CT: Yale University Press, 2013).

Alder, Ken, *The Measure of All Things: The Seven-Year Odyssey and Hidden Error That Transformed the World* (New York: Free Press, 2002).

Allain, Yves-Marie, *Voyages et survie des plantes au temps de la voile* (Paris: Champflour, 2000).

Allen, D. E. 'Parkinson, Sydney (d. 1771)', in *Oxford Dictionary of National Biography* (Oxford: Oxford University Press, 2004); online edn, www.oxforddnb .com/view/article/21377 [accessed 22 April 2016].

Anderson, R. G. W., Marjorie Lancaster, Arthur G. MacGregor and Luke Syson (eds.), *Enlightening the British: Knowledge, Discovery and the Museum in the Eighteenth Century* (London: British Museum Press, 2003).

Bailyn, Bernard and Patricia L. Denault (eds.), *Soundings in Atlantic History. Latent Structures and Intellectual Currents, 1500–1830* (Cambridge, MA and London: Harvard University Press, 2009).

Bandau, A., M. Dorigny and R. von Mallinckrodt (eds.), *Les mondes coloniaux à Paris au XVIIIe siècle: circulation et enchevêtrement des savoirs* (Paris: Editions Karthala, 2010).

Banks, Kenneth, 'Communications and "imperial overstretch": lessons from the eighteenth-century French Atlantic', *French Colonial History* 6 (2005): 17–32.

Banks, R. E. R., B. Elliott, J. G. Hawkes, D. King-Hele and G. L. Lucas (eds.), *Sir Joseph Banks: A Global Perspective* (Richmond: Royal Botanic Gardens, Kew, 1994).

Barker, Hannah, *The Business of Women. Female Enterprise and Urban Development in Northern England, 1760–1830* (Oxford: Oxford University Press, 2006).

Barrier, Janine, Monique Mosser and Che Bing Chiu, *Aux jardins de Cathay: L'imaginaire Anglo-Chinois en Occident. William Chambers* (Paris and Besançon: Éditions de l'Imprimeur, 2004).

Barry, Jonathan, 'Publicity and the public good: presenting medicine in eighteenth-century Bristol', in Bynum and Porter, *Medical Fringe and Medical Orthodoxy*, pp. 29–39.

Batey, Mavis, *Alexander Pope. The Poet and the Landscape* (London: Barn Elms, 1999).

Batsaki, Yota, Sarah Burke Cahalan and Anatole Tchikine (eds.), *The Botany of Empire in the Long Eighteenth Century* (Washington, DC: Dumbarton Oaks, 2016).

Baxter, Denise Amy and Meredith Martin (eds.), *Architectural Space in Eighteenth-Century Europe: Constructing Identities and Interiors* (Ashgate: Farnham, 2010).

Bayly, C. A., *Empire and Information. Intelligence Gathering and Social Communication in India, 1780–1870* (Cambridge: Cambridge University Press, 1996).

Beaglehole, J. C., *The Life of Captain James Cook* (Palo Alto, CA: Stanford University Press, 1974).

Beamont, William, *The History and House of Orford*, ed. H. Wells (Warrington, 1997).

Beaurepaire, Pierre-Yves, and Pierrick Pourchasse (eds.), *Les circulations internationales en Europe, années 1680 – années 1780* (Rennes: Presses Universitaires de Rennes, 2010).

Belhoste, Bruno, *Paris Savant. Parcours et Rencontres au Temps des Lumières* (Paris: Armand Colin, 2012).

Bell, David, *The Cult of the Nation in France: Inventing Nationalism, 1680–1800* (Cambridge, MA and London: Harvard University Press, 2001).

Bending, Stephen, 'Every man is naturally an antiquarian: Francis Grose and polite antiquities', *Art History* 24 (2002): 520–530.

Benedict, Barbara M., *Curiosity. A Cultural History of Early Modern Inquiry* (Chicago, IL and London: University of Chicago Press, 2001).

Benjamin, Marina, 'Elbow room: women writers on science, 1790–1840', in Benjamin, *Science and Sensibility*, pp. 27–59.

(ed.), *Science and Sensibility. Gender and Scientific Enquiry, 1780–1945* (Oxford and Cambridge, MA: Blackwell, 1991).

Berg, Maxine, 'From imitation to invention: creating commodities in eighteenth-century Britain', *Economic History Review* 55 (2002): 1–30.

Luxury and Pleasure in Eighteenth-Century Britain (Oxford: Oxford University Press, 2005).

and Christine Bruland (eds.), *Technological Revolutions in Europe. Historical Perspectives* (Northampton, MA: Edward Elgar, 1998).

and Elizabeth Eger (eds.), *Luxury in the Eighteenth Century: Debates, Desires and Delectable Goods* (Basingstoke: Palgrave, 2003).

and Helen Clifford (eds.), *Consumers and Luxury. Consumer Culture in Europe 1650–1850* (Manchester: Manchester University Press, 1999).

Bermingham, Ann, 'Elegant females and gentleman connoisseurs. The commerce in culture and self image in eighteenth-century England', in Bermingham and Brewer, Consumption of Culture, pp. 489–513.

Learning to Draw. Studies in the Cultural History of a Polite and Useful Art (New Haven, CT: Yale University Press, 2000).

and John Brewer (eds.), *The Consumption of Culture 1600–1800. Image, Object, Text* (London and New York: Routledge, 1995).

Bernard, Bruno (ed.), *Portés par l'air du temps: les voyages du capitaine Baudin* (Brussels: Éditions de l'Université de Bruxelles, 2010).

Berry, Helen, 'Polite consumption: shopping in eighteenth-century England', *Transactions of the Royal Historical Society* 12 (2002): 375–394.

Bewell, Alan, '"On the banks of the South Sea": botany and sexual controversy in the late eighteenth century', in Miller and Reill, *Visions of Empire*, pp. 173–193.

Bhattacharya, Nandini, 'Family jewels: George Colman's Inkle and Yarico and connoisseurship', *Eighteenth-Century Studies* 34 (2001): 207–226.

Bianchi, Marina, 'In the name of the tulip. Why speculation?' in Berg and Clifford (eds.), *Consumers and Luxury*, pp. 88–102.

Black, Jeremy, *Natural and Necessary Enemies: Anglo-French Relations in the Eighteenth Century* (London: Duckworth, 1986).

Bleichmar, Daniela, 'Learning to look: visual expertise across art and science in eighteenth-century France', *Eighteenth-Century Studies* 46(1) (2012): 85–111.

Visible Empire. Botanical Expeditions and Visual Culture in the Hispanic Enlightenment (Chicago, IL: University of Chicago Press, 2012).

Bloch, Jean, 'Discourses of female education in the writings of eighteenth-century French women', in Knott and Taylor, *Women, Gender and Enlightenment*, pp. 243–258.

Blunt, Wilfred, *The Art of Botanical Illustration* (3rd edn, London: Collins, 1955).

The Compleat Naturalist: A Life of Linnaeus (Princeton, NJ, 2001 [1971]).

Bödeker, Hans Erich and Lieselotte Steinbrügge (eds.), *Conceptualising Women in Enlightenment Thought/Conceptualiser la femme dans la pensée des Lumières* (Berlin: Arno Spitz, 2001).

Bolufer Peruga, Mónica, 'Introduction: gender and the reasoning mind', in Knott and Taylor, *Women, Gender and Enlightenment*, pp. 189–194.

Bonneuil, Christophe, 'Mendelism, plant breeding and experimental cultures: agriculture and the development of genetics in France', *Journal of the History of Biology* 39 (2006): 281–308.

Bourde, André J., *Agronomie et Agronomes en France au XVIIIe Siècle* (2 vols, Paris: SEVPEN, 1967).

The Influence of England on the French Agronomes 1750–1789 (Cambridge: Cambridge University Press, 1953).

Bourdieu, Pierre, *The Logic of Practice*, trans. Richard Nice (Cambridge: Polity, 1990).

Bourguet, Marie-Noëlle and Christophe Bonneuil (eds.), *De l'inventaire du monde à la mise en valeur du globe. Botanique et colonisation (fin XVIIe siècle – début XXe siècle)* (Paris: Société française d'histoire d'outre-mer, 1999).

B. Lepetit, D. Nordman and M. Sinarellis, *L'invention scientifique de la Méditerannée: Egypte, Morée, Algérie* (Paris: EHESS, 1998).

Bourret, Michele, Chrystel Fau and Gérard Garriga (eds.), *Parcours. Parcs et Jardins des Hauts-de-Seine* (Nanterre: Conseil Général des Hauts-de-Seine, 1996).

Bowen, H. V., *The Business of Empire. The East India Company and Imperial Britain, 1756–1833* (Cambridge: Cambridge University Press, 2005).

Brears, Peter, 'Commercial museums of eighteenth-century Cumbria. The Crosthwaite, Hutton and Todhunter collections', *Journal of the History of Collections* 4 (1992): 107–126.

Bret, Patrice, 'Les "Indes" en Méditerranée? L'utopie tropicale d'un jardinier des Lumières et la maîtrise agricole du territoire', in Bourguet and Bonneuil, *De l'inventaire du monde*, pp. 65–89.

Brewer, John, *The Pleasures of the Imagination. English Culture in the Eighteenth Century* (Chicago, IL: University of Chicago Press, 1997).

and Ann Bermingham (eds.), *Culture of Consumption: Image, Object, Text* (London and New York: Routledge, 1995).

and Roy Porter (eds.), *Consumption and the World of Goods* (London and New York: Routledge, 1993).

Briost, Pascal, *Espaces maritimes au XVIIIe siècle* (Paris: Atlande, 1997).

Brockliss, L. W. B., *Calvet's Web. Enlightenment and the Republic of Letters in Eighteenth-Century France* (Oxford: Oxford University Press, 2002).

Brockway, Lucile H., *Science and Colonial Expansion. The Role of the British Royal Botanic Gardens* (New York: Academic Press, 1979).

Broman, Thomas, 'The Habermasian public sphere and "science *in* the Enlightenment"', *History of Science* 36 (1998): 123–149.

Brown, Michael, *Performing Medicine: Medical Culture and Identity in Provincial England, c. 1760–1850* (Manchester: Manchester University Press, 2011).

Browne, Janet, 'Botany for gentlemen. Erasmus Darwin and *The Loves of the Plants*', *Isis* 80 (1989): 593–621.

Brygoo, Édouard-Raoul, 'Jardins royaux et princier, en France au XVIIe siècle: Montpellier 1593, Paris 1635, Blois 1636', in Fischer, *Le jardin entre science et représentation*, pp. 65–78.

Bungener, Patrick, 'La botanique au service de l'agriculture. L'exemple des savants genevois', in Robin *et al.*, *Histoire et Agronomie*, pp. 284–301.

'Horace-Bénédict de Saussure (1740–1799), cet illustre inconnu...', *Saussurea. Journal de la Société botanique de Genève* 32 (2002): 61–66.

Burgos, Glenn E. and Daniel J. Kevles, 'Plants as intellectual property. American practice, law, and policy in world context', *Osiris*, 2nd ser. (1992): 75–104.

Burke, Peter and Roy Porter (eds.), *Language, Self and Society. A Social History of Language* (Oxford and Cambridge: Polity, 1991).

Burte, Jean-Noël (ed.), *Le Bon Jardinier. Encyclopédie horticole* (153rd edn, 3 vols, Paris: La Maison Rustique, 1992).

Butel, Paul, *L'économie française au XVIIIe siècle* (Paris: SEDES, 1993).

Bynum, W. F. and Roy Porter (eds.), *Medical Fringe and Medical Orthodoxy, 1750–1850* (London, Sydney and Wolfeboro: Croom Helm, 1987).

Caillé, Jacques, *Un savant montpelliérain: le professeur Auguste Broussonet (1781–1807)* (Paris: A. Pedone, 1972).

Camp, Pannill, *The First Frame. Theatre Space in Enlightenment France* (Cambridge: Cambridge University Press, 2014).

Campbell Orr, Clarissa, 'Aristocratic feminism, the learned governess, and the Republic of Letters', in Knott and Taylor, *Women, Gender and Enlightenment*, pp. 306–325.

Cannadine, David (ed.), *Empire, the Sea and Global History: Britain's Maritime World, c. 1763–c. 1840* (New York: Palgrave Macmillan, 2007).

Cantor, Geoffrey, 'Emanuel Mendes da Costa: constructing a career in science', in Vigne and Littleton, *From Strangers to Citizens*, pp. 230–236.

Caradonna, Jeremy L., *The Enlightenment in Practice. Academic Prize Contests and Intellectual Culture in France, 1670–1794* (Ithaca, NY and London: Cornell University Press, 2012).

Carr, Rosalind, 'A polite and enlightened London?', *The Historical Journal* 59(2) (2016): 623–634.

Carter, G. A., *Warrington and the Mid-Mersey Valley* (Didsbury: E.J. Morten, 1971).

Carter, H. B., *Sir Joseph Banks, 1743–1820* (London: British Museum (Natural History), 1988).

Cayla, J. M., *Histoire des arts et métiers et des corporations ouvrières de la Ville de Paris. Depuis le temps le plus reculés jusqu'à nos jours* (Paris, 1853).

Chappey, Jean-Luc, *Des naturalistes en Révolution. les procès-verbaux de la Société d'Histoire Naturelle de Paris* (Paris: Éditions du CTHS, 2010).

La Société des Observateurs de l'Homme: des anthropologues au temps de Bonaparte (1799–1804) (Paris: Société des Études Robespierristes, 2002).

Chapron, Emmanuelle, 'Du bon usage des recommandations: lettres et voyageurs au XVIIIe siècle', in Beaurepaire and Pourchasse, *Les circulations internationales*.

Chaussinand-Nogaret, Guy, *The French Nobility in the Eighteenth Century. From Feudalism to Enlightenment*, trans. William Doyle (Cambridge: Cambridge University Press, 1985 [1976]).

Clark, William, Jan Golinski and Simon Schaffer (eds.), *The Sciences in Enlightened Europe* (Chicago, IL and London: University of Chicago Press, 1999).

Clifton, Richard 'The Blaikie list. A catalogue of plants in the Jardin du Roi, Paris, 1777', *Gerinaceae Group Associated Note* 38 (2008): 1–50.

Cohen, Deborah and Maura O'Connor (eds.), *Comparison and History. Europe in Cross-National Perspective* (New York, 2004).

Cohen, Michèle, *Fashioning Masculinity: National Identity and Language in the Eighteenth Century* (London: Routledge, 1996).

Colley, Linda, *Britons. Forging the Nation 1707–1837* (2nd edn, New Haven, CT: Yale University Press, 2005).

Coltman, Viccy, *Classical Sculpture and the Culture of Collecting in Britain since 1760* (Oxford: Oxford University Press, 2009).

Colton, Judith, 'Merlin's Cave and Queen Caroline: Garden Art as Political Propaganda', *Eighteenth-Century Studies* 10 (1976): 1–20.

Conan, Michel, 'Histoire des jardins', in Burte, *Le Bon Jardinier*, pp. 970–979.

Connell, David, 'The Grand Tour of William and Winifred Constable 1769–1791', in Hall and Hall, Burton Constable Hall.

Cook, Alexandra, *Jean-Jacques Rousseau and Botany. The Salutary Science*, SVEC 12 (Oxford: Voltaire Foundation, 2012).

'Le pluralisme taxonomique de Jean-Jacques Rousseau', in Jaquier and Léchot, *Rousseau Botaniste*.

'Rousseau and the languages of music and botany', in Dauphin, *Musique et Langage chez Rousseau*, pp. 75–87.

Cook, Harold J., *Matters of Exchange. Commerce, Medicine, and Science in the Dutch Golden Age* (New Haven, CT: Yale University Press, 2007).

Cooper, Alix, 'Homes and households', in Park and Daston, *Cambridge History of Science*.

'Picturing nature: gender and the politics of natural-historical description in eighteenth-century Gdansk/Danzig', *Journal for Eighteenth Century Studies* 36(4) (2013): 519–529.

Coquery, Natacha (ed.), *La boutique et la ville. Commerces, commerçants, espaces et clientèles XVIe–XXe siècle* (Tours: Publication de l'Université François Rabelais, 2000). ·

    L'hôtel aristocratique. Le marché du luxe à Paris au XVIIIe siècle (Paris: Publications de la Sorbonne, 1998).

Corvol, Andrée (ed.), *La nature en révolution 1750–1800* (Paris: L'Harmattan, 1993).

    and Isabelle Richefort (eds.), *Nature, environnement et paysage: l'héritage du 18e siècle. Guide des recherches archivistiques et bibliographiques* (Paris: L'Harmattan, 1995).

Cosgrove, Denis (ed.), *Mappings* (London: Reaktion Books, 1999).

Courpotin, François, 'De la boutique sur rue au magasin: construction et amenagement', in Coquery (ed.), *La boutique et la ville*, pp. 315–337.

Credland, Arthur G., 'Introduction' to Hall and Hall (eds.), *Burton Constable Hall*, pp. 7–9.

Crosland, Maurice, 'Anglo-Continental scientific relations, c. 1780–c. 1820, with special reference to the correspondence of Sir Joseph Banks', in R. E. R. Banks *et al.* (eds.), *Sir Joseph Banks*, pp. 13–22.

    Scientific Institutions and Practice in France and Britain, c. 1700–c. 1870 (Aldershot: Ashgate, 2007).

Crouzet, François, *Britain Ascendant. Comparative Studies in Franco-British Economic History* (Cambridge and Paris: Cambridge University Press and Éditions de la Maison des Sciences de l'Homme, 1990).

Crow, Thomas E., *Painters and Public Life in Eighteenth-Century Paris* (New Haven, CT and London: Yale University Press, 1985).

Crowston, Clare, *Credit, Fashion, Sex: Economies of Regard in Old Regime France* (Durham, NC: Duke University Press, 2013).

    'From school to workshop. Pre-training and apprenticeship in Old Regime France', in de Munck *et al.*, *Learning on the Shop Floor*, pp. 46–62.

Cubitt, Geoffrey (ed.), *Imagining Nations* (Manchester and New York: Manchester University Press, 1998).

Cunningham, Andrew, 'The culture of gardens', in Jardine *et al.*, Cultures of Natural History, pp. 38–56.

Cust, L. H., 'Fairfield, Charles (c. 1759–1804)', rev. Jill Springall, *Oxford Dictionary of National Biography* (Oxford: Oxford University Press, 2004).

Cutting, Rose Marie, *John and William Bartram, William Byrd II and St John de Crèvecoeur: A Reference Guide* (Boston, MA: G. K. Hall & Co., 1976).

Damodaran, Vinita, Anna Winterbottom and Alan Lester (eds.), *The East India Company and the Natural World* (London: Palgrave MacMillan, 2014).

Darnton, Robert, *The Great Cat Massacre and Other Episodes in French Cultural History* (London: Penguin, 2001 [1984]).

Daston, Lorraine, 'Nationalism and scientific neutrality under Napoleon', in Frängsmyr, *Solomon's House Revisited*, pp. 95–119.

    and Elizabeth Lunbeck (eds.), *Histories of Scientific Observation* (Chicago, IL, 2011).

    and Gianna Pomata (eds.), *The Faces of Nature in Enlightenment Europe* (Berlin: BWV-Berliner Wissenschafts-Verlag: 2003).

and Katharine Park, *Wonders and the Order of Nature, 1150–1750* (New York: Zone Books, 1997).

Dauphin, Claude (ed.), *Musique et Langage chez Rousseau*, SVEC 8 (Oxford: Voltaire Foundation, 2004).

Davidoff, Leonora and Catherine Hall, *Family Fortunes: Men and Women of the English Middle Class* (rev. edn, London and New York: Routledge, 2002).

Davies, Martin, 'Knowledge (explicit and implicit): philosophical aspects', in Smelser and Baltes, International Encyclopedia of the Social and Behavioural Sciences, vol. 12, pp. 8126–8132.

de Beer, Gavin, *The Sciences Were Never at War* (London: Thomas Nelson and Sons, 1960).

de Jouvenel, Bertrand, 'The Republic of Science', in *The Logic of Personal Knowledge: Essays presented to Michael Polyani on his 70th Birthday, 11 March 1961* (Glencoe, 1961), pp. 131–141.

Delaporte, François, *Nature's Second Kingdom. Explorations of Plant Vegetality in the Eighteenth Century*, trans. Arthur Goldhammer (Cambridge, MA and London: MIT Press, 1982 [1979]).

Delbourgo, James and Nicholas Dew (eds.), *Science and Empire in the Atlantic World* (New York and Abingdon: Routledge, 2007).

De Marchi, N. and M. Schabas (eds.), *Oeconomies in the Age of Newton*: annual supplement to vol. 35, History of Political Economy (Durham, NC: Duke University Press, 2004).

de Munck, Bert, Steven L. Kaplan and Hugo Soly (eds.), *Learning on the Shop Floor. Historical Perspectives on Apprenticeship* (New York and Oxford: Berghahn Books, 2007).

des Cars, Jean, *Malesherbes. Gentilhomme des Lumières* (Perrin, 2012).

de Vilmorin, Sosthène, 'La famille ANDRIEUX de Sainte-Anne de Guadaloupe', *Généalogie et Histoire de la Caraïbe Bulletin* 80 (March 1996), pp. 1568–1569, www.ghcaraibe.org/bul/ghc080/p1568.html [accessed 28 July 2015].

Dhombres, Nicole and Jean Dhombres, *Naissance d'un Nouveau Pouvoir: Sciences et Savants en France 1793–1824* (Paris: Éditions Payot, 1989).

Dietz, Bettina and Thomas Nutz, 'Collections curieuses: the aesthetics of curiosity and elite lifestyle in eighteenth-century Paris', *Eighteenth-Century Life* 29 (2005): 44–75.

Drayton, Richard, 'À l'école des Français: les sciences et le deuxième empire britannique (1783–1830)', *Revue Française d'Histoire d'Outre-Mer* 86(322–323) (1999): 91–118.

'Maritime networks and the making of knowledge', in Cannadine, *Empire, the Sea and Global History*, pp. 72–82.

Nature's Government. Science, Imperial Britain, and the 'Improvement' of the World (New Haven, CT and London: Yale University Press, 2000).

Drouin, Jean-Marc, 'L'histoire naturelle: problèmes scientifiques et engouement mondain', in Corvol and Rochefort, *Nature, environnement et paysage*, pp. 19–27.

Duris, Pascal, *Linné et la France, 1780–1850* (Geneva: Droz, 1993).

Duthrie, Ruth E., 'English florists' societies and feasts in the seventeenth and first half of the eighteenth centuries', *Garden History* 10 (1982): 17–35.

Dziembowski, Edmond, *Un Nouveau Patriotisme Française, 1750–1770. La France Face à la Puissance Anglaise à l'Époque de la Guerre de Sept Ans*, SVEC 365 (Oxford: Voltaire Foundation, 1998).

Easterby-Smith, Sarah, 'On diplomacy and botanical gifts: France, Mysore and Mauritius in 1788', in Batsaki *et al.*, *Botany of Empire*, pp. 191–209.

'Reputation in a box. Objects, communication and trust in late eighteenth-century botanical networks', *History of Science* 53(2) (2015): 180–208.

'Selling beautiful knowledge: amateurship, botany and the market-place in late eighteenth-century France', *Journal for Eighteenth Century Studies* 36(4) (2013): 531–543.

and Emily Senior, 'The cultural production of natural knowledge: contexts, terms, themes', *Journal for Eighteenth Century Studies* 36(4) (2012): 471–476.

Eger, Elizabeth, *Bluestockings: Women of Reason from Enlightenment to Romanticism* (Basingstoke: Palgrave Macmillan, 2010).

'Luxury, industry and charity: Bluestocking culture displayed', in Berg and Eger (eds.), *Luxury in the Eighteenth Century*, pp. 190–206.

'"The noblest commerce": conversation and community in the Bluestocking circle', in Knott and Taylor, *Women, Gender and Enlightenment*, pp. 288–305.

Egerton, Judy, 'Boyle, Dorothy, countess of Burlington (1699–1758)', in *Oxford Dictionary of National Biography* (Oxford University Press, 2004); online edn, January 2008, www.oxforddnb.com/view/article/66564 [accessed 22 April 2016].

Farber, Paul Lawrence, *Finding Order in Nature: The Naturalist Tradition from Linnaeus to E. O. Wilson* (Baltimore, MD, 2000).

Ferloni, Julia, *Lapérouse. Voyage autour du monde* (Paris: Éditions de Conti, 2005).

Findlen, Paula, *Possessing Nature. Museums, Collecting, and Scientific Culture in Early Modern Italy* (Berkeley and Los Angeles, CA and London: University of California Press, 1994).

Finn, Margot, *The Character of Credit. Personal Debt in English Culture, 1740–1914* (Cambridge: Cambridge University Press, 2003).

Fischer, Jean-Louis (ed.), *Le jardin entre science et représentation* (Paris: Éditions du CTHS, 1999).

Foote, Yolanda, 'Mendes da Costa, Emanuel (1717–1791)', in *Oxford Dictionary of National Biography*, (Oxford: Oxford University Press, 2004); online edn, January 2008: http://www.oxforddnb.com/view/article/6374 [accessed 14 April 2016].

Foucault, Michel, *The Order of Things. An Archaeology of the Human Sciences* (London and New York: Routledge, 1970 [1966]).

Fox, Robert, *The Savant and the State. Science and Cultural Politics in Nineteenth-Century France* (Baltimore, MD: Johns Hopkins University Press, 2012).

Foxhall, Katherine, *Australian Voyages: Convicts, Emigrants, Surgeons and the Sea, c. 1815–1860* (Manchester: Manchester University Press, 2011).

France, Peter, *Politeness and its Discontents. Problems in French Classical Culture* (Cambridge: Cambridge University Press, 1992).

Frängsmyr, Tore (ed.), *Solomon's House Revisited. The Organization and Institutionalization of Science*, Nobel Symposium 75 (Canton, MA: Science History Publications and The Nobel Foundation, 1990).

Franklin, Alfred, *Dictionnaire historique des arts, métiers et professions exercés dans Paris depuis le treizième siècle* (Paris and Leipzig: H. Welter, 1906).

French, H. R., 'The search for the "middle sort of people" in England, 1600–1800', *The Historical Journal* 43(1) (2000): 277–293.

Fuchs, Hans Peter, 'Histoire de la botanique en Valais: I. 1593–1900', *Bulletin Murithienne* 106 (1988): 119–168.

Fulton, John F., 'The Warrington Academy, 1757–1786, and its influence upon medicine and science', *Bulletin of the Institute of the History of Medicine* 1 (1933): 50–80.

Fyfe, Aileen and Bernard Lightman (eds.), *Science in the Marketplace: Nineteenth-Century Sites and Experiences* (Chicago, IL: University of Chicago Press, 2007).

Gascoigne, John, *The Enlightenment and the Origins of European Australia* (Cambridge and New York: Cambridge University Press, 2002).

Joseph Banks and the English Enlightenment. Useful Knowledge and Polite Culture (Cambridge: Cambridge University Press, 1994).

Science in the Service of Empire: Joseph Banks, the British State and the uses of Science in the Age of Revolution (Cambridge: Cambridge University Press, 1998).

Gay, Hannah and Anne Barrett, 'Should the cobbler stick to his last? Silvanus Phillips Thompson and the making of a scientific career', *British Journal for the History of Science* 35 (2002): 151–186.

Gaziello, Catherine, *L'expédition de Lapérouse, 1785–1788: république Française aux voyages de Cook* (Paris, 1984).

George, Sam, *Botany, Sexuality and Women's Writing, 1760–1830. From Modest Shoot to Forward Plant* (Manchester: Manchester University Press, 2007).

'Linnaeus in letters and the cultivation of the female mind: "Botany in an English dress"', *British Journal for Eighteenth-Century Studies* 28 (2005): 1–18.

Gillespie, Richard, 'Ballooning in France and Britain, 1783–1786: aerostation and adventurism', *Isis* 75 (1984): 249–268.

Gillispie, Charles Coulston, (ed.), *Dictionary of Scientific Biography* (18 vols, New York: Scribner, 1970).

Science and Polity in France at the End of the Old Regime (Princeton, NJ: Princeton University Press, 1980).

Science and Polity in France: The Revolutionary and Napoleonic Years (Princeton, NJ: Princeton University Press, 2004).

Gilmartin, Philip M., 'On the origins of heterostyly in *Primula*', *New Phytologist* 208 (2015): 39–51.

Gladstone, Jo, '"New world of English words": John Ray, FRS, the dialect protagonist, in the context of his times (1658–1691)', in Burke and Porter, *Language, Self and Society*, pp. 115–153.

Glaisyer, Natasha, *The Culture of Commerce in England, 1660–1720* (Woodbridge: Royal Historical Society, 2006).

Goffman, Erving, *Behavior in Public Places. Notes on the Social Organization of Gatherings* (New York and London: The Free Press and Collier-MacMillan Ltd, 1963).

Goldgar, Anne, *Impolite Learning. Conduct and Community in the Republic of Letters, 1680–1750* (New Haven, CT and London: Yale University Press, 1995).

'Nature as art: the case of the tulip', in Smith and Findlen, *Merchants and Marvels*, pp. 324–346.

Tulipmania. Money, Honour, and Knowledge in the Dutch Golden Age (Chicago, IL and London: University of Chicago Press, 2007).

Goldsmith, Elizabeth C. and Dena Goodman (eds.), *Going Public. Women and Publishing in Early Modern France* (Ithaca, NY and London: Cornell University Press, 1995).

Golinski, Jan, *Making Natural Knowledge* (2nd edn, Chicago, IL and London: University of Chicago Press, 2005).

Science as Public Culture: Chemistry and Enlightenment in Britain, 1760–1820 (Cambridge: Cambridge University Press, 1992).

Goodman, Dena, *The Republic of Letters: A Cultural History of the French Enlightenment* (Ithaca, NY and London: Cornell University Press, 1994).

'Suzanne Necker's mélanges: gender, writing, and publicity', in Goldsmith and Goodman, *Going Public*, pp. 210–223.

Goodwin, Edward, 'Thoughts on the question, "Whether a Patent, or a Public Premium, is the more eligible Mode of encouraging useful Inventions?"', *Gentleman's Magazine* 56 (1786): 26–27.

Goody, Jack, *The Culture of Flowers* (Cambridge: Cambridge University Press, 2003).

Gornick, Vivian and Barbara Moran (eds.), *Women in Sexist Society: Studies in Power and Powerlessness* (New York: Basic Books, 1971).

Grand-Carteret, Jean, *Les Almanachs Français* (Paris: J. Alisie et Cie, 1896).

Granovetter, Mark S., 'Economic action and social structure: the problem of embeddedness', *American Journal of Sociology* 91(3) (1985): 481–510.

'The strength of weak ties', *American Journal of Sociology* 78(6) (1973): 1360–1380.

Grant, Florence, 'Mechanical experiments as moral exercise in the education of George III', *British Journal for the History of Science* 48(2) (2015): 195–212.

Greig, Hannah, *The Beau Monde: Fashionable Society in Georgian London* (Oxford: Oxford University Press, 2013).

Grieve, Hilda, *A Transatlantic Gardening Friendship, 1694–1777*, Kenneth Newton Memorial Lecture, 1980 (Essex: Historical Association, 1981).

Grove, Richard H., *Green Imperialism: Colonial Expansion, Tropical Island Edens and the Origins of Environmentalism, 1600–1860* (Cambridge: Cambridge University Press, 1995).

Guichard, Charlotte, *Les amateurs d'art à Paris au XVIIIe siècle* (Seyssel: Champ Vallon, 2008).

Hafter, Daryl M. (ed.), *European Women and Preindustrial Craft* (Bloomington, IN: Indiana University Press, 1995).

'Women who wove in the eighteenth-century silk industry of Lyon', in Hafter, European Women and Preindustrial Craft, pp. 42–64.

Hagstrom, W. O., *The Scientific Community* (New York, 1965).

Hahn, Roger, *The Anatomy of a Scientific Institution: The Paris Academy of Sciences, 1666–1803* (Berkeley, CA and London, University of California Press, 1971).

Hall, Elisabeth, 'The cabinet of scientific instruments', in Hall and Hall (eds.), *Burton Constable Hall*, pp. 25–33.

Hall, Ivan and Elisabeth Hall (eds.), *Burton Constable Hall* (Hull: Hull City Museums and Art Galleries and Hutton Press, 1991).

Hancock, David, 'The trouble with networks: managing the Scots' early modern Madeira trade', *Business History Review* 79(3) (2005): 467–491.

Harrison, Carole E., 'Projections of the revolutionary nation: French expeditions in the Pacific, 1791–1803', *Osiris* 24 (2009): 33–52.

Harrison, Mark, 'Science and the British Empire', *Isis* 96 (2005): 56–63.

Harvey, John H., *Early Gardening Catalogues. With Complete Reprints of Lists and Accounts of the Sixteenth to Nineteenth Centuries* (London and Chichester: Phillimore, 1972).

Early Horticultural Catalogues: A Checklist of Trade Catalogues Issued by Firms of Nurserymen and Seedsmen in Great Britain and Ireland down to the Year 1850 (Bath: University of Bath, 1973).

Early Nurserymen. With Reprints of Documents and Lists (London: Phillimore, 1974).

Harvey, Karen, 'Gender, space and modernity in eighteenth-century England: a place called sex', *History Workshop Journal* 51 (2001): 159–180.

Hayden, Ruth, *Mrs Delany. Her Life and her Flowers* (London: British Museum, 1980).

Haynes, Clare, 'A "natural exhibitioner": Sir Ashton Lever and his *Holosphusikon*', *British Journal for Eighteenth-Century Studies* 24 (2001): 1–14.

Headrick, Daniel, *When Information Came of Age. Technologies of Knowledge in the Age of Reason and Revolution, 1700–1850* (Oxford and New York: Oxford University Press, 2000).

Henrey, Blanche, *British Botanical and Horticultural Literature before 1800* (3 vols, London: Oxford University Press, 1975).

Heuzé, Gustav, *Les Vilmorin (1746–1899)* (Paris: Libraire Agricole de la Maison Rustique, 1899).

Heywood, Colin, *The Development of the French Economy, 1750–1914* (Macmillan: Basingstoke and London, 1992).

Hilaire-Pérez, Liliane, 'Diderot's views on artists' and inventors' rights: invention, imitation and reputation', *British Journal for the History of Science* 35 (2002): 129–150.

L'Invention Technique au Siècle des Lumières (Paris: Éditions Albin Michel, 2000).

Hobhouse, Penelope, *Plants in Garden History. An Illustrated History of Plants and their Influences on Garden Styles – From Ancient Egypt to the Present Day* (London: Pavilion, 2004).

Hodacs, Hanna, 'Linnaeans outdoors: the transformative nature of studying nature "on the move" and "outside"', *British Journal for the History of Science* 44(2) (2011): 183–209.

Hooper-Greenhill, Eileen, *Museums and the Shaping of Knowledge* (London and New York: Routledge, 1992).

Hoquet, Thierry (ed.), *Les Fondaments de la Botanique. Linné et la Classification des Plantes* (Paris: Vuibert, 2005).

Horner, Frank, *The French Reconnaissance: Baudin in Australia, 1801–1803* (Melbourne: Melbourne University Press, 1987).

Imbs, Paul (ed.), *Trésor de la Langue Française. Dictionnaire de la Langue du XIXe et du XXe siècle (1789–1960)* (Paris: Éditions du CNRS, 1978).

Jackson Williams, Kelsey, 'Training the virtuoso: John Aubrey's education and early life', *The Seventeenth Century* 27(2) (2012): 157–182.

Jangoux, Michel, 'L'expédition aux Antilles de La Belle Angélique (1796–1798)', in Bernard, *Portés par l'air du temps*, pp. 41–50.

Jaquier, Claire and Timothée Léchot (eds.), *Rousseau Botaniste, Je Vais Devenir Plante Moi-Même* (Fleurier and Pontarlier: Éditions du Belvédère, 2012).

Jardine, N., J. A. Secord and E. C. Spary (eds.), *Cultures of Natural History* (Cambridge: Cambridge University Press, 1996).

Jarrell, Richard A. 'Masson, Francis', in *Dictionary of Canadian Biography* (vol. 5, University of Toronto/Université Laval, 2003–), http://www.biographi.ca/en/bio/masson_francis_5E.html [accessed 30 August 2016].

Jarvis, Robin, *Romantic Writing and Pedestrian Travel* (Basingstoke: Macmillan, 1987).

Jasanoff, Maya, *Edge of Empire. Lives, Culture, and Conquest in the East, 1750–1850* (New York: Alfred A. Knopf, 2005).

Johns, Adrian, 'Natural history as print culture', in Jardine *et al.*, *Cultures of Natural History*, pp. 106–124.

Jones, Colin, 'The great chain of buying: medical advertisement, the bourgeois public sphere, and the origins of the French Revolution', *American Historical Review* 101 (1996): 13–40.

The Great Nation: France from Louis XV to Napoleon (London: Penguin, 2002).

The Smile Revolution in Eighteenth Century Paris (Oxford: Oxford University Press, 2014).

Juliet Carey and Emily Richardson (eds.), *The Saint-Aubin 'Livre de caricatures'. Drawing Satire in Eighteenth-Century Paris* SVEC 6 (Oxford: Voltaire Foundation, 2012).

Jones, Peter M., *Agricultural Enlightenment. Knowledge, Technology, and Nature, 1750–1840* (Oxford: Oxford University Press, 2016).

Industrial Enlightenment. Science, Technology and Culture in Birmingham and the West Midlands, 1760–1820 (Manchester and New York: Manchester University Press, 2008).

Jordanova, Ludmilla, 'Science and nationhood. Cultures of imagined communities', in Cubitt, *Imagining Nations*, pp. 192–211.

Joyaux, François, *La Rose, une Passion Française (1778–1914)* (Brussels: Étidions Complexe, 2001).

Joyce, Patrick, 'What is the social in social history?', *Past and Present* 206 (2010): 213–248.

Kaplan, Steven L., *La fin des corporations* (Paris: Fayard, 2001).

'Social classification and representation in the corporate world of eighteenth-century France' in Kaplan and Koepp (eds.), *Work in France*, pp. 176–228.

and Cynthia J. Koepp (eds.), *Work in France. Representations, Meaning, Organization, and Practice* (Ithaca, NY and London: Cornell University Press, 1986).

Kelly, Gary, *Women, Writing and Revolution 1790–1827* (Oxford and New York: Oxford University Press, 1993).

Kelly, Jason M., *The Society of Dilettanti. Archaeology and Identity in the British Enlightenment* (New Haven, CT: Yale University Press, 2009).

Kevles, Daniel J., 'Patents, protections, and privileges: the establishment of intellectual property in animals and plants', *Isis* 98 (2007): 323–331.

Killingray, David, 'Introduction. Imperial seas: cultural exchange and commerce in the British Empire, 1780–1900', in Killingray *et al.*, *Maritime Empires*, pp. 1–12.

Margarette Lincoln and Nigel Rigby (eds.), *Maritime Empires: British Imperial Maritime Trade in the Nineteenth Century* (Woodbridge: Boydell Press, 2004).

Kingsbury, Pamela Denman, 'Boyle, Richard, third earl of Burlington and fourth earl of Cork (1694–1753)', in *Oxford Dictionary of National Biography* (Oxford University Press, 2004); online edn, January 2008, www.oxforddnb.com/view/article/3136 [accessed 22 April 2016].

Klein, Lawrence E., 'An artisan in polite culture: Thomas Parsons, stone carver, of Bath, 1744–1813', *Huntington Library Quarterly* 75(1) (2012): 27–51.

'Politeness and the interpretation of the British eighteenth century', *The Historical Journal* 45(4) (2002): 869–898.

'Politeness for plebes: consumption and social identity in early eighteenth-century England', in Brewer and Bermingham, *Culture of Consumption*, pp. 362–383.

Shaftesbury and the Culture of Politeness. Moral Discourse and Cultural Politics in Early Eighteenth-Century England (Cambridge: Cambridge University Press, 1994).

(ed.), *Shaftesbury: Characteristics of Men, Manners, Opinions and Times* (Cambridge: Cambridge University Press, 2000).

Klein, Ursula and E. C. Spary, 'Introduction: why materials?', in Klein and Spary, *Materials and Expertise*, pp. 1–24.

(eds.), *Materials and Expertise in Early Modern Europe: Between Market and Laboratory* (Chicago, IL: University of Chicago Press, 2010).

Kleingeld, Pauline, 'Six varieties of cosmopolitanism in late eighteenth-century Germany', *Journal of the History of Ideas* 60 (1999): 505–524.

Knott, Sarah and Barbara Taylor (eds.), *Women, Gender and Enlightenment, 1650–1850* (Basingstoke: Palgrave Macmillan, 2005).

Koerner, Lisbet, *Linnaeus: Nature and Nation* (Cambridge, MA: Harvard University Press, 1999).

Kury, Lorelai, 'Les instructions de voyage dans les expéditions scientifiques françaises (1750–1830)', *Revue d'Histoire des Sciences* 58:1 (1998): 65–91.

Lacour, Pierre-Yves, 'La place des colonies dans les collections d'histoire naturelle 1789–1804', in Bandau, Dorigny and von Mallinckrodt, *Les mondes coloniaux à Paris*, pp. 49–73.

La République Naturaliste. Collections d'Histoire Naturelle et Révolution Française (1789–1804) (Paris: Publications scientifiques du Muséum national d'Histoire naturelle, 2014).

Laird, Mark, *The Flowering of the Landscape Garden. English Pleasure Grounds 1720–1800* (Philadelphia, PA: University of Pennsylvania Press, 1999).

The Formal Garden. Traditions of Art and Nature (London: Thames and Hudson, 1992).

and Alicia Weisberg-Roberts (eds.), *Mrs Delany and Her Circle* (New Haven, CT and London: Yale University Press, 2009).

Laissus, Yves, 'Les cabinets d'histoire naturelle', in Taton, *Enseignement et diffusion des sciences*, pp. 659–712.

'Le Jardin du Roi', in Taton, *Enseignement et diffusion des sciences*, pp. 287–341.

Landes, Joan, *Women and the Public Sphere in the Age of French Revolution* (Ithaca, NY and London: Cornell University Press, 1988).

Langford, Paul, *A Polite and Commercial People* (Oxford: Oxford University Press, 1992).

Larrère, Catherine, 'Women, republicanism and the growth of commerce', in van Gelderen and Skinner, *Republicanism*, vol. 2.

Latour, Bruno, 'On recalling ANT', in Law and Hassard, *Actor Network Theory and After*, pp. 15–25.

Reassembling the Social: An Introduction to Actor-Network-Theory (Oxford: Oxford University Press, 2005).

Science in Action: How to Follow Scientists and Engineers through Society (Cambridge, MA: Harvard University Press, 1987).

'Visualization and cognition: thinking with eyes and hands', *Knowledge and Society: Studies in the Sociology of Culture Past and Present* 6 (1986): 1–40.

Law, John, 'After ANT: complexity, naming and topology', in Law and Hassard, *Actor Network Theory and After*, pp. 1–44.

'On the methods of long-distance control: vessels, navigation and the Portuguese route to India', in Law, *Power, Action and Belief*, pp. 234–263.

(ed.), *Power, Action and Belief. A New Sociology of Knowledge?* (London: Routledge and Kegan Paul, 1986).

and John Hassard (eds.), *Actor Network Theory and After* (Oxford: Blackwell, 1999).

Lawrence, G. H. M. (ed.), *Adanson: The Bicentennial of Michel Adanson's "Familles des Plantes"* (Pittsburgh, PA: Hunt Botanical Library, 1963–1964).

Letouzey, Yvonne, *Le Jardin des Plantes à la Croisée des Chemins avec André Thouin, 1747–1824* (Paris: MNHN, 1989).

Levine, Joseph M., *The Battle of the Books. History and Literature in the Augustan Age* (Ithaca, NY and London: Cornell University Press, 1991).

Lilti, Antoine, *The World of Salons: Sociability and Worldliness in Eighteenth-Century Paris* (abridged version of *Le Monde des Salons* [2005], trans. Lydia G. Cochrane; Oxford: Oxford University Press, 2015).

Lincoln, Margarette (ed.), *Science and Exploration in the Pacific: European Voyages to the Southern Oceans in the Eighteenth Century* (Woodbridge: Boydell Press, 1998).

Lipkowitz, Elise S., 'Seized natural history collections and the redefinition of scientific cosmopolitanism in the era of the French Revolution', *British Journal for the History of Science*, 47(1) (2014): 15–41.

Livesey, James, 'Botany and provincial enlightenment in Montpellier: Antoine Banal père et fils, 1750–1800', *History of Science* 43 (2005): 57–76.

Livingstone, David, *Putting Science in its Place: Geographies of Scientific Knowledge* (Chicago, IL and London: University of Chicago Press, 2003).

Locklin, Nancy, *Women's Work and Identity in Eighteenth-Century Brittany* (Aldershot and Burlington, VT: Ashgate, 2007).

Longstaffe-Gowan, Todd, *The London Town Garden, 1740–1840* (New Haven, CT and London: Yale University Press, 2001).

Lux, David S. and Harold J. Cook, 'Closed circles or open networks?: Communicating at a distance during the scientific revolution', *History of Science* 36 (1998): 179–211.

Lynn, Michael R., *Popular Science and Public Opinion in Eighteenth-Century France* (Manchester and New York: Manchester University Press, 2006).

Lyon, John and Philip R. Sloan (eds.), *From Natural History to the History of Nature: Readings from Buffon and His Critics* (Notre Dame, IN and London, 1981).

Lyons, Henry, *The Royal Society, 1660–1940: A History of its Administration under its Charters* (Cambridge: Cambridge University Press, 1944).

MacGregor, Arthur, *Curiosity and Enlightenment. Collectors and Collections from the Sixteenth to the Nineteenth Century* (New Haven, CT and London: Yale University Press, 2007).

Mack, Phyllis, 'Religion, feminism and the problem of agency', in Knott and Taylor, *Women, Gender and Enlightenment*, pp. 434–459.

Mackay, David, *In the Wake of Cook: Exploration, Science and Empire, 1780–1801* (New York: St Martin's Press, 1985).

MacLeod, Christine, *Inventing the Industrial Revolution. The English Patent System 1660–1800* (Cambridge: Cambridge University Press, 1988).

Mallinson, Jonathan (ed.), *The Eighteenth Century Now: Boundaries and Perspectives*, SVEC 10 (Oxford: Voltaire Foundation, 2005).

Mattelart, Armand, 'Mapping modernity: utopia and communications networks', in Cosgrove, *Mappings*, pp. 169–192.

Maza, Sarah, *The Myth of the French Bourgeoisie: An Essay on the Social Imaginary, 1750–1850* (Boston, MA: Harvard University Press, 2003).

McAleer, John, 'A young slip of botany': botanical networks, the South Atlantic, and Britain's maritime worlds, c. 1790–1810', *Journal of Global History* 11(1) (2016): 24–43.

McClellan, James E. III, and François Regourd, *The Colonial Machine: French Science and Overseas Expansion in the Old Regime* (Turnhout: Brepolis, 2011).

Meredith, Margaret O., 'Friendship and knowledge: correspondence and communication in northern trans-Atlantic natural history, 1780–1815', in Schaffer *et al.*, *The Brokered World*, pp. 151–191.

Metzner, Paul, *Cresecendo of the Virtuoso. Spectacle, Skill, and Self-Promotion in Paris during the Age of Revolution* (Berkeley and Los Angeles, CA and London: University of California Press, 1998).

Meyerowitz, Elliot M., David R. Smyth and John L. Bowman, 'Abnormal flowers and pattern formation in floral development', *Development* 106 (1989): 209–217.

Miller, David Philip, 'Joseph Banks, empire, and "centres of calculation" in late Hanoverian London', in Miller and Reill, *Visions of Empire*, pp. 21–37.

and Peter Hans Reill (eds.), *Visions of Empire. Voyages, Botany and Representations of Nature* (Cambridge and New York: Cambridge University Press, 1996).

Miller, Mary Ashburn, *A Natural History of Revolution. Violence and Nature in the French Revolutionary Imagination, 1789–1794* (Ithaca, NY and London: Cornell University Press, 2011).

Mokyr, Joel, *The Gifts of Athena. Historical Origins of the Knowledge Economy* (Princeton, NJ and Oxford: Princeton University Press, 2002).

Moreau, François, 'Le voyageur français et les étrangers étrangers: bilan d'études sur le siècle des Lumières', in Mallinson, *The Eighteenth Century Now*, pp. 148–159.

Morieux, Renaud, *Une Mer Pour Deux Royaumes. La Manche, Frontière Franco-Anglaise XVIIe-XVIIIe Siècles* (Rennes: Presses Universitaires de Rennes, 2008).

Mowl, Timothy, *William Kent. Architect, Designer, Opportunist* (London: Pimlico, 2007).

Mukerji, Chandra, 'Dominion, demonstration and domination. Religious doctrine, territorial politics, and French plant collection', in Schiebinger and Swan, *Colonial Botany*, pp. 19–33.

Musgrave, Toby, *The Head Gardeners: Forgotten Heroes of Horticulture* (London: Aurum, 2007).

Myers, Sylvia, *The Bluestocking Circle: Women, Friendship, and the Life of the Mind in Eighteenth-Century England* (Oxford: Clarendon, 1990).

Nelson, Amy, 'Is there an international solution to IP protection for plants?', *The George Washington International Law Review* 37 (2005): 997–1030.

Nelson, E. Charles, 'The Dublin Florists' Club in the mid eighteenth century', *Garden History* 10 (1982): 142–148.

Nochlin, Linda, 'Why have there been no great women artists?', in Gornick and Moran, *Women in Sexist Society*.

Nussbaum, Martha C., 'Kant and stoic cosmopolitanism', *The Journal of Political Philosophy* 5 (1997): 1–25.

Nye, Mary Jo, 'On gentlemen, science and the state. A commentary on sessions III and IV', in Frängsmyr, *Solomon's House Revisited*, pp. 322–330.

Ogborn, Miles, 'Writing travels: power, knowledge and ritual on the English East India Company's early voyages', *Transactions of the Institute of British Geographers* 27 (2002): 155–171.

Ogée, Frédéric (ed.), *Better in France? The Circulation of Ideas across the Channel in the Eighteenth Century* (Lewisburg, PA: Bucknell University Press, 2005).

O'Neill, Jean, 'Peter Collinson's copies of Philip Miller's *Dictionary* in the National Library of Wales', *Archives of Natural History* 20 (1993): 373–380.

Outram, Dorinda, *The Enlightenment* (Cambridge: Cambridge University Press, 1995).

Pace, Clare, 'Virtuoso to connoisseur. Some seventeenth-century responses to the visual arts', *The Seventeenth Century* 2 (1987): 166–188.

Park, Katharine and Lorraine Daston (eds.), *The Cambridge History of Science, vol. 3: Early Modern Science* (Cambridge: Cambridge University Press, 2008).

Parrish, Susan Scott, *American Curiosity: Cultures of Natural History in the Colonial British Atlantic World* (Chapel Hill, NC: University of North Carolina Press, 2006).

Parsons, Christopher M., *Cultivating a New France: Knowledge, Empire and Environment in French North America, 1600–1760* (Philadelphia, PA: University of Pennsylvania Press, 2017).

and Kathleen S. Murphy, 'Ecosystems under sail: specimen transport in the eighteenth-century French and British Atlantics', *Early American Studies* 10(3) (2012): 503–529.

Perovic, Sanja, *The Calendar in Revolutionary France: Perceptions of Time in Literature, Culture, Politics* (Cambridge: Cambridge University Press, 2012).

Pomian, Krzysztof, *Collectionneurs, amateurs et curieux. Paris, Venise: XVIe-XVIIIe siècle* (Paris: Gallimard, 1987).

Collectors and Curiosities: Paris and Venice, 1500–1800, trans. Elizabeth Wiles-Portier (Cambridge: Polity, 1990).

Quérard, Joseph-Marie, *La France Littéraire, ou Dictionnaire Bibliographique des Savants, Historiens et Gens de Lettres de la France*, vol. 5 (Paris: Didot, 1833).

Rabier, Christelle (ed.), *Fields of Expertise. A Comparative History of Expert Procedures in Paris and London, 1600 to the Present* (Newcastle-upon-Tyne: Cambridge Scholars, 2007).

Raj, Kapil, *Relocating Modern Science. Circulation and the Construction of Scientific Knowledge in South Asia and Europe. Seventeenth to Nineteenth Centuries* (Basingstoke: Palgrave Macmillan, 2007).

Rappaport, Rhoda, 'Malesherbes, Chrétien-Guillaume de Lamoignon de', in Gillispie, *Dictionary of Scientific Biography*, vol. ix, pp. 53–54.

Reddy, William, *The Navigation of Feeling: A Framework for the History of Emotion* (Cambridge: Cambridge University Press, 2001).

Redford, Bruce, *Dilettanti. The Antic and the Antique in Eighteenth-Century England* (Los Angeles, CA: Getty Publications, 2008).

Rediker, Marcus, *Between the Devil and the Deep Blue Sea: Merchant Seamen, Pirates and the Anglo-American Maritime World, 1700–1750* (Cambridge: Cambridge University Press, 1987).

Rey, Alan (ed.), *Dictionnaire historique de la langue Française* (Paris: Dictionnaires Le Robert, 2006).

Robbins, William J. and Mary Christine Howson, 'André Michaux's New Jersey garden and Pierre Paul Saunier, journeyman gardener', *Proceedings of the American Philosophical Society* 102(4) (1958): 351–370.

Roberts, Lissa, '"*Le centre de toutes choses*": constructing and managing central-ization on the Isle de France', *History of Science* 52(3) (2014): 319–342.

'The circulation of knowledge in early modern Europe: embodiment, mobility, learning and knowing', *History of Technology* 31 (2012): 47–68.

Robin, Paul, Jean-Paul Aeschlimann and Christian Feller (eds.), *Histoire et Agronomie: Entre Ruptures et Durée* (Paris: IRD Éditions, 2007).

Roche, Daniel, *France in the Enlightenment*, trans. Arthur Goldhammer (Cambridge, MA and London: Harvard University Press, 2000 [1993]).

A History of Everyday Things: The Birth of Consumption in France, 1600–1800, trans. Brian Pearce (Cambridge: Cambridge University Press, 2000).

'Natural history in the academies', in Jardine *et al.*, *Cultures of Natural History*, pp. 127–144.

Roger, Jacques, *Buffon. A Life in Natural History*, trans. Sarah Lucille Bonnefoi, ed. L. Pearce Williams (Ithaca, NY and London: Cornell University Press, 1997).

Rogers, Pat, 'Bentinck, Margaret Cavendish [Lady Margaret Cavendish Harley], duchess of Portland (1715–1785)', in *Oxford Dictionary of National Biography* (Oxford: Oxford University Press, 2004); online edn, October 2006: http://www.oxforddnb.com/view/article/40752 [accessed 14 April 2016].

Romano, Antonella and Stéphane Van Damme, 'Penser les savoirs au large (XVIe-XVIIIe siècles)', in Romano and Van Damme, 'Sciences et villes-mondes', pp. 7–18.

(eds.), 'Sciences et villes-mondes, XVIe–XVIIIe siècles', *Revue d'Histoire Moderne et Contemporaine* 55(2) (2008).

Rousseau, G. S. and David Haycock, 'The Jew of Crane Court: Emanuel Mendes da Costa (1717–91), natural history and natural excess', *History of Science* xxxvii (2000): 127–170.

Safier, Neil, 'A courier between empires: Hipólito da Costa and the Atlantic World', in Bailyn and Denault, *Soundings in Atlantic History*, pp. 265–293.

Measuring the New World. Enlightenment Science and South America (Chicago, IL: University of Chicago Press, 2008).

Sairo, Annie, *Language and Letters of the Bluestocking Network. Sociolinguistic Issues in Eighteenth-Century Epistolary English* (Société Néophilologique: Helsinki, 2009).

Sankey, Margaret, Peter Cowley and Jean Fornasiero (eds.), 'The Baudin expedition 1800–1804: texts, contexts and subtexts', special issue of *Australian Journal of French Studies* 41(2) (2004).

Sarrut, Germain, *Biographie des hommes du jour* (6 vols, Paris: Pilout, 1838).

Saunier, Pierre Yves, *Transnational History* (Basingstoke: Palgrave Macmillan, 2013).

Schaffer, Simon, 'The consuming flame: electrical showmen and Tory mystics in the world of goods', in Brewer and Porter, *Consumption*, pp. 489–526.

Lissa Roberts, Kapil Raj and James Delbourgo (eds.), *The Brokered World. Go-Betweens and Global Intelligence* (Sagamore Beach, MA: Science History Publications, 2009).

Schiebinger, Londa, *Plants and Empire. Colonial Bioprospecting in the Atlantic World* (Cambridge, MA and London: Harvard University Press, 2004).

'The private life of plants. Sexual politics in Carl Linnaeus and Erasmus Darwin', in Benjamin, *Science and Sensibility*, pp. 121–143.

and Claudia Swan (eds.), *Colonial Botany. Science, Commerce and Politics in the Early Modern World* (Philadelphia, PA: University of Pennsylvania Press, 2005).

Schlereth, Thomas J., *The Cosmopolitan Ideal in Enlightenment Thought. Its Form and Function in the Ideas of Franklin, Hume, and Voltaire, 1694–1790* (Notre Dame, IN and London: University of Notre Dame Press, 1977).

Schnapper, Antoine, *Le Géant, la licorne et la tulipe. Collections et collectionneurs dans la France du XVIIIe siècle* (Paris: Flammarion, 1988).

Schumpeter, Elizabeth Boody, *English Overseas Trade Statistics, 1697–1808* (Oxford: Clarendon Press, 1960).

Scott, Katie, 'Art and industry. A contradictory union: authors, rights and copyrights during the *Consulat*', *Journal of Design History* 31 (2000): 1–20.

'Authorship, the Académie, and the market in early modern France', *Oxford Art Journal* 21 (1998): 29–41.

Secord, Anne, 'Artisan botany', in Jardine *et al.*, *Cultures of Natural History*, pp. 378–393.

'Corresponding interests: artisans and gentlemen in nineteenth-century natural history', *British Journal for the History of Science* 27 (1994): 383–408.

'Science in the pub: artisan botanists in early nineteenth-century Lancashire', *History of Science* 32 (1994): 269–315.

Secord, James, A. 'How scientific conversation became shop talk', in Fyfe and Lightman, *Science in the Marketplace*.

'Knowledge in transit', *Isis* 95 (2004): 654–672.

Shapin, Steven, 'The image of the man of science', in *The Cambridge History of Science, vol. 4: Eighteenth-Century Science* (Cambridge: Cambridge University Press, 2003): 159–183.

'"The mind is its own place": science and solitude in seventeenth-century England', *Science in Context* 4 (1990): 191–218.

'"A scholar and a gentleman": the problematic identity of the scientific practitioner in early modern England', *History of Science* 29 (1991): 279–327.

A Social History of Truth: Civility and Science in Seventeenth-Century England (Chicago, IL and London: University of Chicago Press, 1994).

and Simon Schaffer, *Leviathan and the Air-Pump: Hobbes, Boyle, and the Experimental Life* (Princeton, NJ: Princeton University Press, 1985).

Sheets-Pyenson, Susan, 'Popular science periodicals in Paris and London: the emergence of a low scientific culture, 1820–1875', *Annals of Science* 42 (1985): 549–572.

Shteir, Ann B., 'Blackburne, Anna (bap. 1726, d. 1793)', in *Oxford Dictionary of National Biography* (Oxford, Oxford University Press, 2004); online edn, January 2008, www.oxforddnb.com/view/article/2512 [accessed 14 April 2016].

'Botanical dialogues: Maria Jacson and women's popular science writing in England', *Journal of Eighteenth-Century Studies* 23 (1990): 301–317.

Cultivating Women, Cultivating Science: Flora's Daughters and Botany in England, 1760–1860 (Baltimore, MD: Johns Hopkins University Press, 1996).

'Priscilla Wakefield's natural history books', in Wheeler and Price, *From Linnaeus to Darwin*, pp. 29–36.

'"With matchless Newton now one soars so high": representing women's scientific learnedness in England', in Bödeker and Steinbrügge, *Conceptualising Women in Enlightenment Thought*, pp. 115–128.

Sigrist, René (ed.), *H.-B. de Saussure (1740–1799): Un Regard Sur la Terre* (Geneva: Georg, 2001).

and Patrick Bungener, 'The first botanical gardens in Geneva (c. 1750–1830): private initiative leading science', *Studies in the History of Gardens and Designed Landscapes* 28 (2008).

Sloboda, Stacey, 'Displaying materials: Porcelain and natural history in the Duchess of Portland's museum', *Eighteenth-Century Studies* 43(4) (2010): 455–472.

'Fashioning Bluestocking conversation: Elizabeth Montagu's Chinese room', in Baxter and Martin, *Architectural Space*, pp. 129–148.

Smail, John, 'Credit, risk and honor in eighteenth-century commerce', *Journal of British Studies* 44 (2005): 439–456.

Smelser, N. J. and P. B. Baltes (eds.), *International Encyclopedia of the Social and Behavioural Sciences* (26 vols, Amsterdam: Elsevier Science, 2001).

Smith, Pamela H. and Benjamin Schmidt (eds.), *Making Knowledge in Early Modern Europe. Practices, Objects, and Texts, 1400–1800* (Chicago, IL and London: University of Chicago Press, 2007).

and Paula Findlen (eds.), *Merchants and Marvels: Commerce, Science and Art in Early Modern Europe* (New York: Routledge, 2002).

Sorrenson, Richard, 'The ship as a scientific instrument in the eighteenth century', *Osiris* 11 (1996).

Spary, E. C., 'Forging nature at the republican Muséum', in Daston and Pomata, *The Faces of Nature in Enlightenment Europe*, pp. 163–180.

'The "nature" of Enlightenment', in Jardine *et al.*, *Cultures of Natural History*.

'"Peaches which the patriarchs lacked": Natural history, natural resources, and the natural economy in eighteenth-century France', in De Marchi and Schabas, *Oeconomies in the Age of Newton*, pp. 14–41.

'Scientific symmetries', *History of Science* 42 (2004): 1–46.

Utopia's Garden. French Natural History from Old Regime to Revolution (Chicago, IL and London: University of Chicago Press, 2000).

Stafleu, Frans A., *Linnaeus and the Linnaeans. The Spreading of their Ideas in Systematic Botany, 1735–1789* (Utrecht: A. Oosthoek's Uitgeversmaatscchappj N.V., 1971).

Starbuck, Nicole, *Baudin, Napoleon and the Exploration of Australia* (London: Pickering and Chatto, 2013).

Stemerding, Dirk, *Plants, Animals and Formulae. Natural History in the Light of Latour's Science in Action and Foucault's The Order of Things* (Enschede: University of Twente, 1991).

Stewart, Larry, *The Rise of Public Science: Rhetoric, Technology and Natural Philosophy in Newtonian Britain, 1660–1750* (Cambridge: Cambridge University Press, 1992).

Stobart, Jon, Andrew Hann and Victoria Morgan, *Spaces of Consumption. Leisure and Shopping in the English Town, c. 1680–1830* (London and New York: Routledge, 2007).

Sutton, Geoffrey V., *Science for a Polite Society. Gender, Culture and the Demonstration of Enlightenment* (Boulder, CO and Oxford: Westview Press, 1995).

Swan, Claudia, 'Collecting naturalia in the shadow of the early modern Dutch trade', in Schiebinger and Swan, *Colonial Botany*, pp. 223–236.

Swann, Marjorie, *Curiosities and Texts. The Culture of Collecting in Early Modern England* (Philadelphia, PA: University of Pennsylvania Press, 2001).

Swem, E. G., 'Brothers of the Spade. Correspondence of Peter Collinson, of London, and of John Custis, of Williamsburg, Virginia, 1734–1746', *Proceedings of the American Antiquarian Society* 58 (1949): 17–190.

Taton, René (ed.), *Enseignement et diffusion des sciences en France au XVIIIe siècle* (Paris: Hermann, 1964).

Taylor, Patricia, *Thomas Blaikie (1751–1838): The 'Capability' Brown of France* (East Linton: Tuckwell Press, 2001).

Terrall, Mary, *Catching Nature in the Act. Réaumur and the Practice of Natural History in the Eighteenth Century* (Chicago, IL: University of Chicago Press, 2013).

'Frogs on the mantelpiece: the practice of observation in daily life', in Daston and Lunbeck, *Histories of Scientific Observation*.

Teute, Fredrika, 'The loves of the plants; or, the cross-fertilization of science and desire at the end of the eighteenth century', *Huntington Library Quarterly* 63 (2000): 319–345.

Thomson, Ann, Simon Burrows and Edmond Dziembowski (eds.), *Cultural Transfers: France and Britain in the Eighteenth Century*, SVEC 2010(4) (Oxford: Voltaire Foundation, 2010).

Tikanoja, Tuomas, *Transgressing Boundaries: Worldly Conversation, Politeness and Sociability in Ancien Régime France, 1660–1789* (Helskinki: Unigrafia Oy, 2013).

Tombs, Robert and Isabelle Tombs, *That Sweet Enemy: The French and the British from the Sun King to the Present* (London, 2006).

Torrens, Hugh, 'Natural history in eighteenth-century museums in Britain', in Anderson *et al.*, *Enlightening the British*, pp. 81–91.

Toynbee, Paget (ed.), *Strawberry Hill Accounts. A Record of Expenditure in Building, Furnishing &c., Kept by Horace Walpole from 1747 to 1795* (Oxford: Clarendon Press, 1927).

Turner, Katherine, 'Twiss, Richard (1747–1821)', in *Oxford Dictionary of National Biography* (Oxford: Oxford University Press, 2004); online edn, October 2006: www.oxforddnb.com/view/article/27918 [accessed 14 April 2016].

Uglow, Jenny, *The Lunar Men. The Friends Who Made the Future. 1730–1810* (London: Faber, 2002).

Van Damme, Stéphane, *Paris, Capitale Philosophique. De la Fronde à la Révolution* (Paris: Odile Jacob, 2005).

van de Roemer, Bert, 'Neat nature: the relation between nature and art in a Dutch cabinet of curiosities from the early eighteenth century', *History of Science* 42 (2004): 47–84.

van Gelderen, Martin and Quentin Skinner (eds.), *Republicanism: A Shared European Heritage* (2 vols, Cambridge, 2002).

Vanneste, Tijl, *Global Trade and Commercial Networks* (London: Pickering and Chatto, 2011).

Versini, Laurent (ed.), *Diderot. Œuvres* (Paris: Robert Laffont, 1996).

Vickery, Amanda, *The Gentleman's Daughter. Women's Lives in Georgian England* (New Haven, CT and London: Yale University Press, 1998).

'The theory and practice of female accomplishment', in Laird and Weisberg-Roberts, *Mrs Delany and Her Circle*, pp. 94–109.

Vigne, Randolf and Charles Littleton (eds.), *From Strangers to Citizens. The Integration of Immigrant Communities in Britain, Ireland and Colonial America, 1550–1750* (London and Brighton: The Huguenot Society of Great Britain and Sussex Academic Press, 2001).

Walsh, Claire, 'Shopping et tourisme: l'attrait des boutiques parisiennes au XVIIIe siècle', in Coquery, *La boutique et la ville*, pp. 223–237.

Walters, Alice. N, 'Conversation pieces: science and politeness in eighteenth-century England', *History of Science* 35 (1997): 121–154.

Webber, Ronald, *The Early Horticulturalists* (Newton Abbot: David and Charles, 1968).

Werner, Michael and Bénédicte Zimmermann (eds.), *De la Comparaison à l'Histoire Croisée* (Paris: Seuil, 2004).

Wheeler, A. and J. H. Price (eds.), *From Linnaeus to Darwin: Commentaries on the History of Biology and Geology* (London: Society for the History of Natural History, 1985).

White, Paul, 'The purchase of knowledge: James Edward Smith and the Linnaean collections', *Endeavour* 23 (1999): 126–129.

Whitting, Philip D., (ed.), *A History of Hammersmith, Based upon that of Thomas Faulkner* (London: Hammersmith Local History Group, 1965).

Williams, Roger L., *Botanophilia in Eighteenth-Century France: The Spirit of the Enlightenment* (Dordrecht and Boston, MA: Kluwer Academic Publishers, 2001).

Willson, E. J., 'Farming, market and nursery gardening', in Whitting, *History of Hammersmith*, pp. 88–100.

James Lee and the Vineyard Nursery, Hammersmith (London: Hammersmith Local History Group, 1961).

Wilson, Richard and Alan Mackley, *Creating Paradise. The Building of the English Country House, 1660–1880* (London and New York: Hambledon and London, 2000).

Withers, Charles, *Geography, Science and National Identity: Scotland Since 1520* (Cambridge: Cambridge University Press, 2001).

Placing the Enlightenment. Thinking Geographically about the Age of Reason (Chicago, IL and London: Chicago University Press, 2007).

Wood, Henry Trueman, *A History of the Royal Society of Arts* (London: J. Murray, 1913).

Woodcroft, Bennet, *Subject-Matter Index . . . of Patents of Invention, from March 2, 1617 . . . to October 1, 1852* (2 vols, London: George Eyre and William Spottiswoode, 1857).

Wulf, Andrea, *The Brother Gardeners. Botany, Empire and the Birth of an Obsession* (London: William Heinemann, 2008).

Wystrach, V. P., 'Anna Blackburne (1726–1793) – a neglected patroness of natural history', *Journal of the Society for the Bibliography of Natural History* 8 (1977): 148–168.

'Ashton Blackburne's place in American ornithology', *The Auk* 92 (1975): 607–610.

Yeo, Richard, *Notebooks, English Virtuosi, and Early Modern Science* (Chicago, IL: University of Chicago Press, 2014).

UNPUBLISHED PAPERS AND THESES

Coulton, Richard, 'Curiosity, commerce, and conversation in the writing of London horticulturalists during the early eighteenth century', unpublished PhD thesis, Queen Mary University of London (2005).

Easterby-Smith, Sarah, 'Propagating commerce. Plant breeding and market competition in London and Paris, ca. 1770 – ca. 1800', in Gaudillière *et al.*, *Living Properties*.

Gaudillière, Jean-Paul, Daniel J. Kevles and Hans-Jörg Rheinberger (eds.), *Living Properties: Making Knowledge and Controlling Ownership in the History of Biology*, Preprint 382 (Berlin: Max Planck Institute for the History of Science, 2009).

Hollis, Dawn L., 'Re-thinking mountains: ascents, aesthetics, and environment in early modern Europe', unpublished PhD thesis, University of St Andrews (2016).

Horsman, Frank, 'Botanising in Linnaean Britain: a study of Upper Teesdale in northern England', unpublished PhD thesis, University of Durham (1998).

Hubbard, Philippa, 'The art of advertising: Trade cards in eighteenth-century consumer cultures', unpublished PhD thesis, University of Warwick (2009).

Lustig, Abigail J., 'The creation and uses of horticulture in Britain and France in the nineteenth century', unpublished DPhil thesis, University of California, Berkeley (1997).

Murphy, Kathleen S., 'Portals of nature: networks of natural history in eighteenth-century British plantation societies', unpublished PhD thesis, Johns Hopkins University (2007).

Romero-Passerin, Elena, 'Un jardin des lumières. Le *Royal Botanic Garden* d'Édimbourg à la fin du XVIIIe siècle', unpublished MA dissertation, UFR d'Histoire, Université de Paris-Sorbonne (2015).

Traversat, Michel, 'Les pépinières: étude sur les jardins français et sur les jardiniers et les pépiniéristes', unpublished PhD thesis, EHESS, Paris (2001).

WORLD WIDE WEB SOURCES

International Union for the Protection of New Varieties of Plants (UPOV), C(Extr.)/19/2 Rev. 'The Notion of Breeder and Common Knowledge' (9 August 2002), www.upov.int/about/en/key_issues.html [accessed 11 August 2016].

Index